Studies in Computational Intelligence   420

Editor-in-Chief

Prof. Janusz Kacprzyk
Systems Research Institute
Polish Academy of Sciences
ul. Newelska 6
01-447 Warsaw
Poland
E-mail: kacprzyk@ibspan.waw.pl

For further volumes:
http://www.springer.com/series/7092

Joaquim P. Marques de Sá, Luís M.A. Silva,
Jorge M.F. Santos, and Luís A. Alexandre

Minimum Error
Entropy Classification

 Springer

Authors
Joaquim P. Marques de Sá
INEB-Instituto de Engenharia Biomédica
Porto
Portugal

Luís M.A. Silva
Dept. of Mathematics
Univ. de Aveiro
Campus Universitário de Santiago
Aveiro
Portugal

Jorge M.F. Santos
Dept. of Mathematics
ISEP, School of Engineering
Polytechnic of Porto
Porto
Portugal

Luís A. Alexandre
Dept. of Informatics
Univ. Beira Interior
IT - Instituto de Telecomunicações
R. Marquês d'Ávila e Bolama
Covilhã
Portugal

ISSN 1860-949X
ISBN 978-3-642-43742-7
DOI 10.1007/978-3-642-29029-9
Springer Heidelberg New York Dordrecht London

e-ISSN 1860-9503
ISBN 978-3-642-29029-9 (eBook)

Springer is part of Springer Science+Business Media (www.springer.com)

To our wives and children.

Foreword

It is with great pleasure that I write this foreword for two main reasons. First, I know the first author for over 30 years and had the opportunity to interact with all the authors in several occasions when they were piecing together the scientific puzzle that lead to this book. I will like to commend the authors for their hard work and vision because their hypothesis and results coalesced into a piece of focused and highly relevant material that will undoubtedly improve our understanding for the role of information theoretic descriptors in learning theory. Second, because information theoretic learning is one of my preferred areas of research, I read the book avidly and found highly relevant material that increased my knowledge on the subject and provided new directions for my future research work. I am certain that any researcher, graduate student or practitioner in pattern recognition and machine learning will also benefit tremendously from reading the monograph to help build their foundations on information theoretic learning to solve current data processing applications or to seek new topics for research.

When I started my work on Information Theoretic Learning I concentrated on regression/filtering because these problems are simpler to analyze theoretically and with estimators, and avoided the issues that are at the core of this book. I must say that the strategy of focusing the text on minimum entropy error (MEE) criteria for classification is very successful because the problem can be studied in greater depth, the presentation provides a clear view of the challenges and fills a void that existed in the literature. I also appreciated greatly the didactic style of explaining the difficulties and illustrating all the cases with simple but well thought out examples.

The book carefully distinguishes theoretical cost functions from the empirical ones achievable in practice through estimators. This is rather important because it provides future directions for both theoretical studies and also allows practitioners to apply the novel algorithms in real world applications. Surprisingly, it shows experimentally that the smoothing effects of kernels in fact provide better behaved solutions in many cases. As the authors illustrate, there are substantial improvements in the use of MEE and its simplifications versus MSE in many data sets with the same computational complexity. Therefore there are plenty of practical applications for entropy cost functions in classification. The book does not provide all the answers to questions in this research area, however it establishes a productive approach and solid methodological foundations to move forward. So it is very timely and complements admirably the available literature.

Gainesville, Florida,   Jose C. Principe
February 2012

Preface

Information Theoretic Learning has received a good deal of attention in recent years, specifically in what concerns the application of error entropy as a risk functional to be minimized by a learning machine. Theoretical investigation on the inner workings of this Minimum Error Entropy (MEE) concept and also its application to a variety of problems soon brought up interesting theoretical insights on risk functionals and have shown the practical usefulness of MEE.

For learning machines performing a regression task, the MEE approach implies a reduction of the expected information contained in the error, leading to the maximization of the mutual information between the desired target and the output of the machine. This means that learning of the target variable is taking place in such regression machines, an important result justifying the good performance of MEE in such tasks as times series forecast and blind source separation.

In this book, we concentrate on the application of MEE to classifiers, a novel topic when we started our research work on it some eight years ago. As shown in the book, the application of MEE to classification raises difficult and even perplexing issues with no parallel in regression tasks. Our aim is to elucidate these issues, and with a special focus on classifier attainment of the minimum probability of error allowed by the specific function family it implements.

Concomitantly, we present a broad display of classifiers using the MEE approach, from simple to complex, and describe them and their application to a variety of tasks: supervised classification with either continuous or discrete errors, unsupervised and modular classification, classification of symbolic input sequences, and of complex-valued data. Examples, tests, evaluation experiments and comparison with similar machines using classic approaches, complement the descriptions.

The book is aimed at machine learning researchers and practitioners. The reader is assumed to have some familiarity with analysis, probability, and calculus. We remind the reader of some important results, so it should suffice

just to have met the related concepts before. We also assume the reader to
have some basic knowledge of the data classification (or pattern recognition)
area; we do, however, introduce many of the important aspects of data classification in a simple but rigorous way, that we deem sufficient to grasp the
explanations of new concepts and results.

We thank the "FCT — Fundação para a Ciência e a Tecnologia", "INEB
— Instituto de Engenharia Biomédica", and "UBI — Universidade da Beira
Interior", for the financial and logistic support of a large part of the research
work without which this book wouldn't have been possible. A word of grat-
itude is also due to Professor Jose Principe of the CNEL — Computational
NeuroEngineering Laboratory (University of Florida) for helpful suggestions
and encouragements, and to Professor Mark Embrechts for many interesting
discussions.

Finally, a big thank to our wives and children for their encouragement and
tolerance over the years.

Porto,   Joaquim P. Marques de Sá
January 2012   Luís M.A. Silva
                                                             Jorge M.F. Santos
                                                            Luís A. Alexandre

Contents

1 Introduction 1
  1.1 Data Classification 1
  1.2 Data-Based Learning 4
  1.3 Outline of the Book.................................... 9

2 Continuous Risk Functionals 13
  2.1 Classic Risk Functionals 14
    2.1.1 The Mean-Square-Error Risk 14
    2.1.2 The Cross-Entropy Risk........................... 19
  2.2 Risk Functionals Reappraised 21
    2.2.1 The Error Distribution 21
    2.2.2 MSE and CE Risks in Terms of the Error 22
  2.3 The Error-Entropy Risk 26
    2.3.1 EE Risks 26
    2.3.2 EE Risks and Information Theoretic Learning........ 28
    2.3.3 MEE Is Harder for Classification
       than for Regression 29
    2.3.4 The Quest for Minimum Entropy 31
    2.3.5 Risk Functionals and Probability of Error 34

3 MEE with Continuous Errors 41
  3.1 Theoretical and Empirical MEE 42
    3.1.1 Computational Issues 42
    3.1.2 Empirical EE Gradient Descent 44
    3.1.3 Fat Estimation of Error PDF 45
  3.2 The Linear Discriminant 50
    3.2.1 Gaussian Inputs................................. 50
    3.2.2 Consistency and Generalization 55
  3.3 The Continuous-Output Perceptron...................... 59
    3.3.1 Motivational Examples........................... 60
    3.3.2 Theoretical and Empirical MEE Behaviors.......... 67
    3.3.3 The Arctangent Perceptron 71

3.4   The Hypersphere Neuron 80
      3.4.1   Motivational Examples 80
      3.4.2   Theoretical and Empirical MEE in Realistic Datasets . 84
3.5   The Data Splitter 86
3.6   Kernel Smoothing Revisited 88

4   MEE with Discrete Errors 93
4.1   The Data Splitter Setting 94
      4.1.1   Motivational Examples 96
      4.1.2   SEE Critical Points 98
      4.1.3   Empirical SEE Splits 104
      4.1.4   MEE Splits and Classic Splits 110
4.2   The Discrete-Output Perceptron 116

5   EE-Inspired Risks 121
5.1   Zero-Error Density Risk 122
      5.1.1   Motivational Examples 122
      5.1.2   Empirical ZED 124
5.2   A Parameterized Risk 133
      5.2.1   The Exponential Risk Functional 133

6   Applications 139
6.1   MLPs with Error Entropy Risks 139
      6.1.1   Back-Propagation of Error Entropy 140
      6.1.2   The Batch-Sequential Algorithm 149
      6.1.3   Experiments with Real-World Datasets 152
6.2   Recurrent Neural Networks 156
      6.2.1   Real Time Recurrent Learning 157
      6.2.2   Long Short-Term Memory 161
6.3   Complex-Valued Neural Network 170
      6.3.1   Introduction 170
      6.3.2   Single Layer Complex-Valued NN 171
      6.3.3   MMSE Batch Algorithm 172
      6.3.4   MEE Algorithm 173
      6.3.5   Experiments 174
6.4   An Entropic Clustering Algorithm 176
      6.4.1   Introduction 176
      6.4.2   Overview of Clustering Algorithms 177
      6.4.3   Dissimilarity Matrix 178
      6.4.4   The LEGClust Algorithm 179
      6.4.5   Experiments 192
6.5   Task Decomposition and Modular Neural Networks 199
      6.5.1   Modular Neural Networks 199
      6.5.2   Experiments 201
6.6   Decision Trees 204

6.6.1 The MEE Tree Algorithm 205
6.6.2 Application to Real-World Datasets 208

A   Maximum Likelihood
    and Kullback-Leibler Divergence 215
    A.1 Maximum Likelihood 215
    A.2 Kullback-Leibler Divergence 216
    A.3 Equivalence of ML and KL Empirical Estimates 216

B   Properties of Differential Entropy 219
    B.1 Shannon's Entropy 219
    B.2 Rényi's Entropy 220

C   Entropy and Variance of Partitioned PDFs 221

D   Entropy Dependence on the Variance 223
    D.1 Saturating Functions 223
    D.2 PDF Families with Up-Saturating $H(V)$ 225
        D.2.1 The Pearson System 225
        D.2.2 List of PDFs with USF $H(V)$ 229

E   Optimal Parzen Window Estimation 235
    E.1 Parzen Window Estimation 235
    E.2 Optimal Bandwidth and IMSE 238

F   Entropy Estimation 241
    F.1 Integral and Plug-in Estimates 241
    F.2 Integral Estimate of Rényi's Quadratic Entropy 241
    F.3 Plug-in Estimate of Shannon's Entropy 243
    F.4 Plug-in Estimate of Rényi's Entropy 244

References 245

Index 257

Abbreviations and Symbols

Abbreviations

iff	If and only if
i.i.d.	Independent and identically distributed
r.v.	Random variable
std	Standard deviation
s.t.	Such that
w.l.o.g.	Without loss of generality
w.r.t.	With respect to
CDF	Cumulative Distribution Function
CE	Cross-Entropy
EE	Error-Entropy
GI	Gini Index
IG	Information Gain
IMSE	Integrated Mean Square Error
KDE	Kernel Density Estimation (or Estimate)
MCE	Minimum Cross-Entropy
MEE	Minimum Error Entropy
ML	Maximum Likelihood
MLP	Multilayer Perceptron
MSE	Mean Square Error
MMSE	Minimum Mean Square Error
NN	Neural Network (artificial neural network)
PDF	Probability Density Function
PMF	Probability Mass Function
RBF	Radial Basis Function (network)
R_2EE	Rényi's Quadratic Error Entropy
SEE	Shannon's Error Entropy
SVM	Support Vector Machine

Symbols

Sets, variables and functions

A	Random variable; the codomain (target set) of r.v. A. We use X, Y or Z, T, and E, for classifier input, output, target, and error r.v., respectively. We also use capital italic letters to denote sets, namely sets of parameters.
a	Any value of A. We always denote a value by the same letter as the variable it refers to, but in lower case. We also use small italic letters to denote functions and constants, namely n for the number of instances, d for the number of instance dimensions, and c for the number of classes.
$\mathcal{A}_W = \{a_w : X \to T;\ w \in W\}$	Set (family) of parametrized functions.
$A = \{a_1, a_2, \ldots\}$	Set of $a_1, a_2, \ldots$.
$\lvert A \rvert$	Cardinality of set A.
$\Omega = \{\omega_1, \omega_2, \ldots\}$	Set of class labels (text constants) $\omega_1, \omega_2, \ldots$
$]a, b]$	Interval between a and b, excluding a. Also represented as $(a, b]$ in the literature. We similarly use open bracket instead of parenthesis for other intervals.
$\mathbb{C}$	Complex number set.
$\mathbb{N}$	Natural number set.
$\mathbb{R} =]-\infty, +\infty[$	Real number set.
$\mathbb{Z}$	Integer number set.
$\overline{A}$	Complement set of A.
$\lfloor x \rfloor$	The largest integer smaller or equal to x.
$\mathbb{1}_A(x)$	Indicator function ($\mathbb{1}_A(x) = 1$ if $x \in A$ or property A is verified; 0, otherwise).
$G(x)$	Gauss Kernel function.
$\Gamma(x)$	Gamma function.
$\ln(x)$	Natural logarithm of x.
$\mathrm{sgn}(x)$	Sign of x ($\mathrm{sgn}(x) = 1$ if $x \geq 0$; -1, otherwise).
$\tanh(x)$	Hyperbolic tangent of x.
$df(x)/dx, f'(x)$	Derivative of function f w.r.t. x.
$\otimes$	Convolution operator.
L_p	p norm.
$\mathscr{L}^p$	Space of the L_p Lebesgue-integrable functions.

Vectors and Matrices

$\mathbf{x}$	Vector (column vector).
$[x_1 \ x_2 \ \ldots \ x_n]$	Row vector with components $x_1, x_2, \ldots, x_n$.
$\|\mathbf{x}\|$	Norm of $\mathbf{x}$.
$\mathbf{A}$	Matrix.
$\mathbf{x}^T, \mathbf{A}^T$	Transpose of vector $\mathbf{x}$, transpose of matrix $\mathbf{A}$.
$\mathbf{x}^T \mathbf{y}$	Inner product of $\mathbf{x}$ and $\mathbf{y}$.
$\mathbf{A}^{-1}$	Inverse of matrix $\mathbf{A}$.
$|\mathbf{A}|$	Determinant of matrix $\mathbf{A}$.
$\mathbf{I}$	Unit matrix.

Probabilities

$P_A(a)$	Probability of $A = a$, also written as $P(A = a)$. We may also simply write $P(a)$, when no confusion arises as to the r.v. referred to. We also use P to denote discrete probability values. Moreover, P_e denotes the probability of misclassification (error) and P_c the probability of correct classification.
$\hat{P}$	Estimate of P. The caret always means an estimate.
$f_X(x), F_X(x)$	Respectively, PDF and CDF relative to X evaluated at x (we may drop the X when no confusion arises).
$X|t, X|T$	X conditioned on $T = t$; X conditioned on $T = t$, for every $t \in T$.
$P(x|t)$	Probability of $X = x$ conditioned on t (the same as $P(X = x|t)$).
$f_{X|t}(x)$	PDF of $X|t$ evaluated at x (we may drop $X|t$ when no confusion arises).
$X \sim f$	X has PDF (is distributed as) f.
$u(x; a, b)$	Uniform PDF in $[a, b]$ evaluated at x.
$g(x; \mu, \sigma)$	Gaussian (normal) PDF with mean μ and standard deviation σ evaluated at x. The standardized normal PDF is $g(x; 0, 1) \equiv G(x)$.
$\Phi(x)$	Standardized Gaussian (normal) CDF: $\Phi = \int_{-\infty}^{x} g(u; 0, 1) du$.
$g(\mathbf{x}; \mathbf{\mu}, \mathbf{\Sigma})$	Gaussian (normal) multivariate PDF with mean $\mathbf{\mu}$ and covariance $\mathbf{\Sigma}$ evaluated at $\mathbf{x}$.

$\mathbb{E}_{X,Y}[\phi(X,Y,\ldots)]$ Expected value (average, mean) of $\phi(X,Y,\ldots)$, w.r.t. X,Y, i.e., defined in terms of the joint CDF $F_{X,Y}(x,y)$. The subscripts may be dropped when no confusion arises (e.g., $\mathbb{E}_X[X] = \mathbb{E}[X]$).

$\mathbb{E}[X|y], \mathbb{E}[X|Y]$ Expected value of X conditioned on $Y = y$; expected value of X conditioned on $Y = y$, for every $y \in Y$ ($\mathbb{E}[X|Y] = \mathbb{E}_{X,Y}[X|Y]$).

$V[X], V[f]$ Variance of X; Variance of a r.v. with PDF f.

$D_{KL}(p\|q)$ Kulback-Leibler divergence of p relative to q.

$H(X)$ Entropy of r.v. X.

Chapter 1
Introduction

1.1 Data Classification

The advent of automatic data classification can be traced back to the early sixties, namely to a software tool for discriminating abnormal from normal electrocardiograms which was actually in routine use for some time [172]. Two important resources were then available, a theoretical one — Bayesian decision theory —, and a practical one — computational facilities, albeit extremely modest by present day standards.

Bayesian decision theory, with its statistical model-based approach to the classification problem, was a main ingredient in the toolbox of the Pattern Recognition area, when its blossoming started at the mid-sixties. In accordance to this approach, one attempts to find an adequate probabilistic model for the dataset and, taking the model for granted, one then proceeds to using well-known statistical estimation techniques in order to derive object discrimination (decision) rules into classes. Soon, however, researchers and practitioners came to realize the shortcomings of the model-based approach to classification, particularly when in the usual practice "model-based" really meant "Gaussian-distribution-based". Even the pioneers of automatic electrocardiogram diagnostic became soon aware that, for many diagnostic classes of interest, their data was everything less than "Gaussian distributed". The search for so-called "non-parametric" approaches to solve classification problems was then pursued and came to include mathematical devices (machines) such as decision trees, nearest-neighbor classifiers, (artificial) neural networks, and support vector machines. As a matter of fact, these devices are indeed parametrically tuned in order to accomplish their task; we avoid confusion and gain language precision by referring to them (and their underlying methods) as being data-based in opposition to model-based.

Data-based approaches are not exclusively applied to solving classification problems; they are also applied to regression and probability density estimation problems. In all these three types of problems — classification,

J. Marques de Sá et al.: Minimum Error Entropy Classification, SCI 420, pp. 1–11.
springerlink.com   © Springer-Verlag Berlin Heidelberg 2013

regression, and probability density (PDF) estimation — the data-based device has to learn some desired information — respectively, class labeling, functional description, and probability density — from the data.

To formalize the classification problem, we start by assuming that a dataset X_{ds} is available for the inductive design of the classifier: design or training set. The training set X_{ds} can be viewed as an array whose rows correspond to data *objects* (e.g., individual electrocardiograms for the above electrocardiogram classification problem), and whose columns represent object *attributes* (measurements, features). We denote by n the number of objects (also called instances or cases) of X_{ds}. Each instance is represented by an ordered sequence of attributes with d elements x_j from some space X (the input space of the classification system). The attributes can be numerical — and in this case we always assume an underlying real number domain — , or nominal (categorical), say a set B of categories. For the above electrocardiogram classification problem an instance is represented by electrocardiographic signal features (amplitudes and durations of signal waves), measured as real numbers, and by categorical features such as sex ($B =$\{"male", "female"\}).

We will often be dealing with instances characterized solely by numerical attributes; in this case $X_{ds} \subset X = \mathbb{R}^d$, and any instance $x \in X$ is (represented as) an ordered sequence (d-tuple): $x = (x_1, x_2, \ldots, x_j, \ldots, x_d)$. Sometimes we may find it convenient to use vector notation for x, $\mathbf{x} = [x_1 \, x_2 \ldots x_j \ldots x_d]^T$, specifically when vector operations are required; X_{ds} is then represented by an $n \times d$ real matrix.

Any attribute value x_j is a realization value of a random variable (r.v.) X_j, whose codomain is X_j; whether X_j denotes a codomain or a variable will be obvious from the context. Note that X_j may have a single Dirac-δ distribution, in which case X_j is in fact a deterministic variable (a degenerate random variable). We will also denote by X the d-dimensional r.v. whose codomain is X and whose realization values are the d-tuples $x = (x_1, x_2, \ldots, x_j, \ldots, x_d)$; X will be characterized by a joint distribution of the X_j with cumulative distribution function F_X.

Throughout the book all data instances in X_{ds} are assumed as having been obtained by an independent and identically distributed (i.i.d.) sampling process, from a d-dimensional joint probability distribution with cumulative distribution function F_X characterizing a large (perhaps infinite) *population* of instances. For numerical attributes defined in bounded intervals of $\mathbb{R}$ (as the electrocardiographic measurements) one may still use the real line as domain, by assigning zero probability outside the intervals.

When confronted with *unsupervised classification* problems (popularly known as data clustering problems), i.e., when one wants the classification system to find a structuring solution that partitions the data into "meaningful" groups (clusters) according to certain criteria, the X_{ds} set, $X_{ds} = \{x_i = (x_{i1}, x_{i2}, \ldots, x_{ij}, \ldots, x_{id}); \; i = 1, \ldots, n\}$, is all that is required. Data clustering is a somewhat loose type of classification problem, since one may find a variety of solutions (unsupervised classifiers) depending on the clustering properties (therefore, on object similarity criteria) one may impose. Moreover, even though there are validity techniques applicable to gauging the clustering solutions, they can be nonetheless evaluated from different perspectives.

For *supervised classification* problems, besides the X_{ds} set one also requires a set Ω_{ds} of class labels assigned to the data objects by a supervisor. For instance, for the above electrocardiogram classification problem, it is assumed that a supervisor (physician, in this case) labeled each electrocardiogram as either being "normal" or "abnormal". We denote by $\Omega = \{\omega_k; \; k = 1, \ldots, c\}$ the set of $c \in \mathbb{N}$ possible class labels (e.g., $\Omega = \{\text{"normal", "abnormal"}\}$ for the two-class electrocardiogram problem) and we will usually find it convenient to code the labels with numerical values from a set $T = \{t_k; \; k = 1, \ldots, c\} \subset \mathbb{Z}$, using some one-to-one $\Omega \to T$ mapping function (e.g., $\Omega = \{\text{"normal", "abnormal"}\} \to T = \{0, 1\}$). We call T the *target* value set. We then have $T_{ds} = \{t_i \in T; \; i = 1, 2, \ldots, n\}$ as a set of n target values $t_i = t(x_i)$ assigned by some unknown labeling function, $t : X \to T$. The target values are seen as instantiations of a target r.v. also denoted T.

We call supervised classifier, or just classifier, any $X \to T$ mapping implemented by a supervised classification system. Designing a classifier corresponds to picking one function z_w out of a family of functions $\mathcal{Z}_W = \{z_w : X \to T; \; w \in W\}$, through the selection (tuning) of a parameter (either a single or multi-component parameter sequence) w from W. Examples of classifiers are decision trees and neural networks, where $\mathcal{Z}_W$ corresponds to all $X \to T$ mappings that the architecture of these devices implements, w being a particular choice of parameters (respectively, thresholds and weights), and W the parameter space. The classifier output $Z_w = z_w(X)$ is a r.v. whose codomain is a subset of T.

A classifier may implement an $X \to T$ mapping using several "physical" outputs: tree leaves in the case of decision trees, network output neurons in the case of neural networks, etc. In some cases an arbitrarily large number of "physical" outputs may exist; for instance, a decision tree can have an arbitrarily large number of hierarchical levels with a consequent arbitrarily large number of leaves, the "physical" outputs. Mathematically, we are often only interested in characterizing a single $X \to T$ mapping, independently of how many "physical" outputs were used to implement such mapping. The association to class labels can be materialized in various ways. For instance, decision trees have as many outputs as there are tree leaves, whose number is usually larger than c; since each tree leaf represents a single class label, as a consequence each class label may be represented by more than one leaf. Other classifiers, such as many neural networks, have instead c outputs, as many as class labels, with the possible exception of two-class problems for which only one output is needed, usually, coded 1 for one of the classes, and 0 or -1 for its complement. For $c > 2$ it is customary to express both the target values and the outputs as c-dimensional vectors using a 1-of-c coding scheme.

(A coding scheme such as $T = \{1, \ldots, c\}$ turns out to be uncomfortable to deal with.) For instance, for $c = 4$ the classifier would have four outputs, $\{z_1, z_2, z_3, z_4\}$, and class ω_2, say, would be represented by $[0\ 1\ 0\ 0]^T$ or $[-1\ 1\ -1\ -1]^T$; i.e., with z_2 discriminating ω_2 from its complement. We then see that the 1-of-c coding scheme allows to analyze the classification task as a set of dichotomizations, whereby any class is discriminated from its complement. It is not surprising, therefore, that in the following chapters we will concentrate on studying two-class classification problems; the results of the study transpose directly to multi-class problems.

Classifiers, such as several types of neural networks (an example of which are multilayer perceptrons), are *regression-like* machines: the classifier builds a *continuous* approximation to the targets, using some function y_w, which is subsequently thresholded in order to yield numerical values according to the coding scheme. For instance, when neural network outputs are produced by sigmoid units in the $[0, 1]$ range one may set a threshold at 0.5; an output larger than 0.5 is assigned the class label corresponding to $T = 1$ (say, ω_2), otherwise to its complement ($\overline{\omega}_2$), corresponding to $T = 0$. The function y_w can be interpreted in a geometrical sense, with its $X \times y_w(X)$ graph representing a *decision surface* that separates (discriminates) instances from distinct classes. The classifier mapping is $z_w : X \to T$, with $z_w = \theta_w(y_w)$ and θ_w an appropriate thresholding function.

Neural networks and decision trees are used extensively in the present book to illustrate the minimum error entropy approach to supervised classification. There is an abundant literature on theoretical issues, as well as on architectures and design methods of neural networks and decision trees. Relevant books on neural networks are [26,27,142,96]; for decision trees, see [33] and [177]. Supervised and unsupervised classifiers, especially with a focus on Pattern Recognition applications, are presented in detail by [59] and [225]. A main reference on theoretical issues concerning classifier operation and performance is [52].

1.2 Data-Based Learning

For a classifier whose design is based on a set $(X_{ds}, T_{ds}) = \{(x_i, t(x_i)); i = 1, 2, \ldots, n\}$ some type of approximation of the output $z_w(x)$ to $t(x)$, for all $x \in X$, is required. From all possible approximation criteria one can think of there is one criterion that clearly stands out, which addresses the mismatch between observed and predicted class labels: how close is the classifier of achieving the minimum probability of (classification) error, P_e, in $X \times T$? We are then interested in the efficacy of the classifier in the population, $X \times T$.

Figure 1.1 illustrates a data-based classification system implementing $z_w \in \mathcal{Z}_W$, where the parameter w is tuned by a *learning algorithm*, built expressly for the design process alone. The learning algorithm performs an

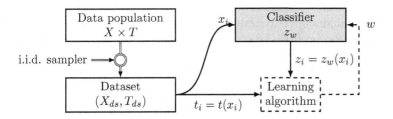

Fig. 1.1 A classification system implementing z_w with w supplied by a learning algorithm.

induction task by picking a *general* classification rule z_w based on some *particular* training set (X_{ds}, T_{ds}). It accomplishes the induction task by reflecting through w the matching (or mismatching) rate between classifier *outputs*, $z_i = z(x_i)$, and *targets*, $t_i = t(x_i)$, for all instances of (X_{ds}, T_{ds}). The earlier mentioned assumption of the training set being obtained by i.i.d. sampling is easily understandable: any violation of this basic assumption will necessarily bias the inductive learning process.

Let us assume that the training algorithm produced w and, therefore, the classifier implements z_w. The probability of error (P_e) of this classifier is:

$$P_e(Z_w) = \mathbb{E}_{X,T}[t(X) \neq z_w(X)] , \qquad (1.1)$$

which can also be expressed as

$$P_e(Z_w) = \mathbb{E}_{X,T}[L(t(X), z_w(X))] = \int_{X \times T} L(t(x), z_w(x)) dF_{X,T}(x, t) \qquad (1.2)$$

with

$$L(t(x), z_w(x)) = \begin{cases} 0, & t(x) = z_w(x) \\ 1, & t(x) \neq z_w(x) \end{cases} . \qquad (1.3)$$

Formula (1.2) expresses P_e as the expected value in $X \times T$ of the function L, called *loss* function (also called cost or error function). For a given $t(X)$ the probability of error depends on the classifier output r.v. $Z_w = z_w(X)$. The expectation is computed with the cumulative distribution function $F_{X,T}(x, t)$ of the joint distribution of the random variables X and T representing the classifier inputs and the targets. The above loss function is simply an indicator function of the misclassified instances: $L(\cdot) = \mathbb{1}_{\{t \neq z_w\}}(\cdot)$.

There are variations around the probability of error formulation, related to the possibility of weighing differently the errors according to the class membership of the instances. We will not pay attention to these variations.

Similar formulations of the learning task can be presented if one is dealing with regression or PDF estimation problems. In other words, one can present a general formulation that encompasses all three learning problems. In this general formulation one has a *learning machine* whose algorithm attempts to achieve a minimum approximation performance measure, with distinct flavors according to the problem type (see e.g., [41]). In the present book our sole concern is with classification problems.

Regarding the approximation performance of classifiers, several important points are to be noted:

1. We usually don't know $F_{X,T}$, which renders impossible the integration in formula (1.2), even in the improbable scenario that other technicalities were easily solvable. In practical real-world problems, where one designs z_w using the n-sized dataset (X_{ds}, T_{ds}), one can only obtain an estimate of P_e, the so-called *empirical error estimate* (or empirical error rate), which is simply the average number of misclassified instances:

$$\hat{P}_e(Z_w) \equiv \hat{P}_e(n) = \frac{1}{n} \sum_{x \in X_{ds}} \mathbb{1}_{\{t \neq z_w\}}(x) \ . \tag{1.4}$$

2. The question then arises of how large the training set size, n, should be such that with a given confidence (say, 95% confidence) the empirical error rate (the *observed error*), $\hat{P}_e$, does not deviate from the probability of error (the *true error*), P_e, more than a fixed tolerance. The answer to this question is not difficult for model-based classifiers: just use the available techniques of statistical estimation (namely, the computation of confidence intervals). For data-based classifiers this *generalization issue* is rather more intricate. In both cases its analysis leads one to using a "functional complexity" of $\mathcal{Z}_W$ adequate to the training set size. Intuitively, too much complexity of $\mathcal{Z}_W$ will tend to produce complex decision surfaces that tightly discriminate the training set instances, but will be unable to perform equally well (generalize) in new independent datasets (the *over-fitting* phenomenon); too less complexity will produce a classifier that generalizes well but with poor performance, failing to extract from the data all possible information (*under-fitting*). The present book doesn't include a discussion on the generalization issue and related issues such as dimensionality reduction and function regularization. Theoretical and practical aspects on these topics are discussed in detail in [52,230,41,227,11,228]. For shorter but well structured and informative surveys, see [30,233]. We will include a few remarks on these aspects when appropriate. We will also make use of the simple generalization measure $|P_e - \hat{P}_e|$, which is small if the classifier generalizes well. Note that generalization has nothing to do with P_e being small; it really means that $\hat{P}_e$ is a good estimate of P_e.

3. For a classifier architecture that implements $\mathcal{Z}_W = \{z_w : X \to T; w \in W\}$ what is really of interest to us is not some z_w but the best of them all. In other words, one requires the learning algorithm to achieve the *minimum* probability of error in $\mathcal{Z}_W$, expressed as follows:

Find w s.t. $P_e(Z_w)$ is minimum.

The best classifier is

$$z_{w^*} = \operatorname*{argmin}_{z_w \in \mathcal{Z}_W} P_e(Z_w) \, . \tag{1.5}$$

The classifier $z_{w^*} : X \to T$, with optimal parameter w^*, is the best one (in the minimum P_e sense) in the family $\mathcal{Z}_W$. We will often denote $P_e(Z_{w^*})$ simply as $\min P_e$, signifying $\min_{\mathcal{Z}_W} P_e$, the minimum probability of error for the functional family allowed by the classifier architecture.

An important aspect concerning the estimates $\hat{P}_e(Z_w) \equiv \hat{P}_e(n)$ produced by a classifier is whether or not they will converge (in some sense) with growing n to $\min P_e$. This *consistency* issue of the learning algorithm will be addressed when appropriate.

4. If one knew the class priors, $P(t_k)$, and the class conditional distributions of the targets, $p(x|t_k)$, with p representing either a PMF or a PDF, one would then be able to determine the best possible classifier based on the Bayes decision theory: just pick the class that maximizes the posterior probability

$$P(t_k|x) = \frac{p(x|t_k)P(t_k)}{p(x)}, \text{ with } p(x) = \sum_{k=1}^{c} p(x|t_k)P(t_k) \, . \tag{1.6}$$

This is the procedure followed by the model-based approach to classification. The best — $P(t_k|x)$ maximizing — classifier is known as the Bayes classifier, z_{Bayes}.

One always has $P_e(Z_w) \geq P_e(Z_{w^*}) \geq P_e(Z_{Bayes})$. Note that there will be function families $\mathcal{Z}_B$ such that $z_{w^*}(\cdot) = z_{Bayes}(\cdot)$ with $w^* \in B$ (e.g., multilayer perceptrons with "enough" hidden neurons are known to have universal functional approximation capabilities); however, one usually will not be sure whether or not $\mathcal{Z}_B$ is implementable by the classification system being used (for multilayer perceptrons, "enough" may not be affordable, among other things because of the generalization issue). We, therefore, will not pursue the task of analyzing the approximation of data-based classifiers to z_{Bayes}.

We also shall not discuss whether z_w is convergent with n (in some sense) to z_{Bayes}, the so-called Bayes-consistency issue, largely dependent on the classification system being used; as a matter of fact, the lack of Bayes-consistency does not preclude the usefulness of a classification system (binary decision trees with impurity decision rules are an example of that). For details on the consistency of classification systems the reader may find useful to consult [52] and [11].

Let us now address the problem of how to find the best classifier z_{w^*}, affordable by the function family $\mathcal{Z}_W$ implemented by the classification system. One could consider using formula (1.4) (with large n so that $\hat{P}_e(n)$ is close to $P_e(Z_{w^*})$) and perform an exhaustive search in some discrete version of the parameter space W. However, except for too simple problems, such a search is impractical; specifically, for both decision trees and multilayer perceptrons such a search is known to be NP complete (see, respectively, [108] and [29]; although the proof in the latter respects to 3-node neural networks, its extension to more complex architectures is obvious). (For the reader unfamiliar with computational complexity theory, we may informally say that "NP complete" refers to a class of optimization problems for which no known efficient way to locate a solution exists.)

For regression-like machines such as neural networks, the practical approach to finding a solution close to z_{w^*} is by employing some sort of optimization algorithm (e.g., gradient descent). There is, however, a requirement of optimization algorithms that precludes using the expectation expressed by formula (1.2): the loss function must be continuous and differentiable. One then needs to work with the continuous output, Y_w, of these machines and, instead of using the discontinuous loss function in expression (1.2), one uses some suitable continuous and differentiable loss function; i.e., instead of attempting to find the minimum probability of error one searches for a solution minimizing the *risk*

$$R_L(Y_w) = \int_{X \times T} L(t(x),\ y_w(x)) dF_{X,T}(x,t) \ , \tag{1.7}$$

which is an expected value of some continuous and differentiable loss function $L(t(x), y_w(x))$, $y_w \in \mathcal{Y}_W$: $R_L(Y_w) = \mathbb{E}_{X,T}[L(t(X), y_w(X))]$, where $y_w(\cdot)$ is now the classifiers's continuous output (preceding the thresholding operation). The risk, also known in the literature as error criterion, expresses an average penalization of deviations of $y_w(\cdot)$ from the target $t(\cdot)$, in accordance to the loss function. For a given machine and data distribution the risk depends, therefore, on the loss function being used.

Similarly to what happened with the probability of error, we now have to deal in practice with an *empirical risk*, computed as:

$$\hat{R}_L(Y_w) = \frac{1}{n} \sum_{x \in X_{ds}} L(t(x),\ y_w(x)) \ . \tag{1.8}$$

Let us assume that, based on the minimization of the empirical risk, we somehow succeeded in obtaining a good estimate of the *minimum risk* (not $\min P_e$) solution; that is, we succeeded in obtaining a good estimate of

$$\hat{R}_L(Y_{w^+}) \quad \text{with} \quad y_{w^+} = \arg \min_{y_w \in \mathcal{Y}_W} \hat{R}_L(Y_w) \ , \tag{1.9}$$

where y_{w^+} denotes the minimum risk classifier; $z_{w^+} = \theta(y_{w^+})$ is not guaranteed to be z_{w^*}, the $\min P_e$ classifier in $\mathcal{Z}_W = \theta(\mathcal{Y}_W)$.

For decision trees, some kind of greedy algorithm is used, that constructs the tree as a top-down sequence of locally optimized decisions at each tree node, until meeting some stop criterion at a tree leaf. For a given tree architecture, the final solution is, as a consequence of the greedy approach, never guaranteed to be close to z_{w^*}. Moreover, it turns out that the empirical error is usually a poor choice of criterion for local decisions; instead, some other type of *risk* criterion is used.

When carrying out risk optimization in regression-like machines, such as multilayer perceptrons, one must not loose sight of the fact that the fundamental reason for attempting to optimize the risk expressed by formula (1.7), instead of the special case of risk expressed by formula (1.2), is the tractability of the former (efficient optimization algorithms for continuous and differentiable loss functions) and the intractability of the latter. That is, we consider $R_L(Y_w)$ for tractability reasons alone, even though our real interest is in $P_e(Z_w)$.

Algorithms for the minimization of $\hat{R}_L(Y_w)$ can be of two sorts: *direct parameter estimation* algorithms; *iterative optimization* algorithms (which may be of a stochastic or deterministic nature). As an example of the first category of algorithms, we have the well-known *normal equations* for solving linear regression problems, based on a statistical estimation of the linear coefficients. The normal equations are directly derived from the minimization of $\hat{R}_L(Y_w)$ by setting $\partial \hat{R}_L(Y_w)/\partial w = 0$. This direct approach can be generalized to more complex regression problems, using kernel tricks (see e.g., [207]). A disadvantage of the direct methods is that one may have to operate with large matrices when solving practical problems, and large matrices raise ill-conditioning and regularization issues which may be difficult to cope with. Iterative optimization algorithms may then be preferred. These follow a stepwise adaptive approach, adjusting the parameter vector at each step in an adequate way that takes into account the first order (and sometimes the second order) derivatives of $\hat{R}_L(Y_w)$. Examples of such *classifier training* (or learning) algorithms are all the gradient descent procedures (such as the well-known Widrow-Hoff adaptive method and the back-propagation training of multilayer perceptrons), as well as the popular conjugate gradient and Levenberg-Marquardt algorithms.

For a given classification problem does a particular choice of loss function, and therefore of $\min R_L(Y_w)$, provide the $P_e(Z_{w^*})$ solution? We call this issue the *classifier problem*: the choice of L leading to a classifier $z_{w^+}(\cdot)$ achieving the same performance as $z_{w^*}(\cdot)$ (note that there may be several such classifiers). We shall see in the following chapter that loss functions do not always perform equally in what concerns this issue, which also motivates the Minimum Error Entropy approach for classification that we introduce.

1.3 Outline of the Book

The Minimum Error Entropy (MEE) concept is introduced in the following Chap. 2, where we start by presenting classic risk functionals and discuss their main features. We then move on to derive risk descriptions in terms of error random variables as a preliminary step to the introduction of error entropies as risks. Risk functionals are then put under the Information Theoretic Learning perspective, advantages and pitfalls on the application of error entropies to classification tasks are identified, and specific entropic properties influencing the MEE approach are discussed and exemplified. Chapter 2 concludes with a discussion that risk functionals do not perform equally in what respects the attainment of $\min P_e$ solutions, and provides concrete evidence to this respect.

Chapter 3 presents theoretical and experimental results regarding the application of the MEE approach to simple classifiers having continuous error distributions, and which are used as building blocks of more sophisticated machines: linear discriminants, perceptrons, hypersphere neurons, and data splitters. Classification examples for artificial and realistic datasets (based on real-world data) are presented. Consistency and generalization issues are studied and discussed. The $\min P_e$ issue is also analyzed in detail both from a theoretical (theoretical MEE) and a practical (empirical MEE) perspective. Regarding the practical perspective and the attainment of $\min P_e$ solutions, the influence of kernel smoothing on error PDF estimates is also scrutinized.

In Chap. 4 we devote our attention to the application of MEE to discrete error distributions, provided by simple classifiers having a threshold activation function. We study with care the case of a data splitter for univariate input discrimination and prove some results regarding the $\min P_e$ issue. Two versions of empirical MEE splits are proposed and experimentally tested both with artificial and real-world data: the kernel-based approach and the resubstitution estimate approach. The latter introduces the concept of MEE splits for decision tree classifiers. This new splitting criteria is carefully compared to the classic splitting criteria. The chapter concludes with the analysis of a discrete-output perceptron, investigating how error entropy critical points relate to $\min P_e$ solutions and then analyzing the special case of bivariate Gaussian two-classs problems.

Chapter 5 introduces two new risk functionals. First, the EE-inspired Zero-Error Density (ZED) risk functional is defined and illustrated both for the discrete and continuous error distribution cases. Its empirical version for the case of continuous error distributions (empirical ZED) is presented. Empirical ZED is then tested in a perceptron learning task and its gradient behavior is compared to the ones of other classic risks. Connections between ZED and the correntropy measure are also discussed. Finally, we present as a generalization of the ZED risk, the Exponential (EXP) risk functional, a parameterized risk sufficiently flexible to emulate a whole range of behaviors, including the ones of ZED and other classic risks.

Chapter 6 describes various types of classifiers applying the MEE concept. Popular classifiers such as multilayer perceptrons and decision trees are studied, as well as more sophisticated types of classifiers, such as recurrent, complex-valued, and modular networks. The latter use unsupervised classification algorithms, an advanced MEE version of which is also described and studied. Besides implementation issues, experimental results obtained with MEE-based classifiers on real-world datasets, are also presented and compared with those obtained using the non-MEE versions of the classifiers.

Chapter 2
Continuous Risk Functionals

As explained in the preceding chapter, the learning algorithm needed to adequately tune a regression-like classifier, based on the information provided by a training set, consists of the minimization of a quantity called risk, whose expression is given by formula (1.7). This formula assigns a number, $R_L(Y_w)$, to a function y_w, i.e., the formula is an instantiation of an $\mathcal{Y}_W = \{y_w\} \to \mathbb{R}$ mapping. Such mapping type (from a set of functions onto a set of numbers) is called a *functional*. The risk functional, expressed in terms of a *continuous and differentiable* loss function $L(t(x), y_w(x))$, is minimized by some algorithm attempting to find a classifier with a probability of error hopefully close to that of $z_{w^*} : \min P_e$. From now on we assume that the class conditional distributions are continuous[1] and, as a consequence, the risk functional can be expressed as

$$R_L(Y) = \sum_{t \in T} P(t) \int_{X|t} L(t(x), y(x)) f_{X|t}(x) dx , \qquad (2.1)$$

where $f_{X|t}(x)$ is the class-conditional density (likelihood) of the data, $P(t)$ is the prior probability of class (with target value) t, and $y \equiv y_w$. We have already pointed out in the preceding chapter that $R_L(Y)$ is an expected value:

$$R_L(Y) = \mathbb{E}_{X,T}[L(t(X), y(X))] = \sum_{t \in T} P(t) \mathbb{E}_{X|t}[L(t, y(X))] . \qquad (2.2)$$

As also mentioned in Chap. 1, since the class-conditional densities are usually unknown, the minimization carried out by learning algorithms is performed on an empirical estimate of (2.1) (also known as *Resubstitution estimate*), $\hat{R}_L(Y)$, expressed by formula (1.8). With mild conditions on L (measurability), the empirical estimate $\hat{R}_L(Y)$ converges to $R_L(Y)$ almost surely as $n \to \infty$.

---

[1] In order to have a density the distribution must be absolutely continuous. However, since we will not consider datasets exhibiting exotic continuous singular distributions, the continuity assumption suffices.

J. Marques de Sá et al.: Minimum Error Entropy Classification, SCI 420, pp. 13–39.
springerlink.com   © Springer-Verlag Berlin Heidelberg 2013

Note that for simplicity of notation we write $R_L(Y)$ and $\hat{R}_L(Y)$ instead of $R_L(Y_w)$ and $\hat{R}_L(Y_w)$. One must keep in mind that the minimization of both risks is with respect to $w \in W$. For a fixed family of output functions y_w, we could as well denote the risk as $R_L(w)$.

We now proceed to surveying the two classic risk functionals that have been almost exclusively used in learning algorithms, before introducing the new risk functionals that are the keystones of the present book.

2.1   Classic Risk Functionals

2.1.1   The Mean-Square-Error Risk

The oldest and still the most popular, continuous and differentiable loss function, is the square-error (SE) function

$$L_{SE}(t(x), y(x)) = (t(x) - y(x))^2 , \qquad (2.3)$$

with corresponding risk functional

$$R_{MSE}(Y) = \sum_T P(t) \int_X (t(x) - y(x))^2 f_{X|t}(x) dx . \qquad (2.4)$$

The empirical estimate of this functional is

$$\hat{R}_{MSE}(Y) = \frac{1}{n} \sum_{i=1}^{n} (t_i - y_i)^2 \qquad (2.5)$$

with $t_i = t(x_i)$ and $y_i = y(x_i) \equiv y_w(x_i)$.

The empirical risk expressed by formula (2.5) corresponds to the well-known mean-square-error (MSE) method introduced by Gauss in the late 18th century as a means of adjusting a function to a set of observations. In the present context the observations are the t_i and we try to fit the $y_w(x_i)$ to the t_i, for a set of predictor values x_i. Formula (2.5) expresses a penalization of the deviations of $y_w(x_i)$ from t_i, according to a square law, therefore emphasizing large deviations between observed and predicted values. The square law (2.3) is a distance measure (for both sequences and functions), and is still the praised measure in regression because of its several important properties and its mathematical tractability.

Let us consider that, to some deterministic data generating process $g(X)$, some noise, $\xi(X)$, is added: $Z = g(X) + \xi(X)$. X and ξ are both random variables. The minimum mean-square-error (MMSE) estimate $Y = f(X)$ of $g(X)$ based on Z and the square-error measure, i.e., the $\min \mathbb{E}[(Z - Y)^2]$ solution, turns out to be the conditional expectation of Z given X: $Y =$

$\mathbb{E}[Z|X]$ (see e.g., [136]). This is the usual regression solution of Z predicted by X. One of the reasons why the MMSE estimate Y is so praised in regression problems, is that it is the optimal one — affords the minimum $\mathbb{E}[L(Z - Y)]$ for a class of convex, symmetric, and unimodal loss functions — when $g(X)$ is linear and X and ξ are Gaussian [208, 88]. Furthermore, when the noise is independent of X and has zero mean, the conditional expectation factors out as $Y = \mathbb{E}[g(X)|X] + \mathbb{E}[\xi(X)|X] = g(X)$. One is then able to retrieve $g(X)$ from Z.

For classification problems the MMSE solution also enjoys important properties. Instead of deriving these properties from the regression setting (applying the above $Z = g(X) + \xi(X)$ model to classification raises mathematical difficulties), they can be derived [83, 185, 26, 252] by first observing that the empirical MSE risk, $\hat{R}_{MSE}$, for a classifier with c target values t_k and outputs y_k is written as

$$\hat{R}_{MSE} = \frac{1}{n} \sum_{k=1}^{c} \sum_{i=1}^{n_k} (t_{ik} - y_k(x_i))^2 , \tag{2.6}$$

where n_k is the number of instances of class ω_k and each y_k depends on the parameter vector w. For $n \to \infty$, and after some mathematical manipulations, one obtains:

$$\hat{R}_{MSE} \xrightarrow[n \to \infty]{} R_{MSE} = \sum_{k=1}^{c} \int_{X|T} (\mathbb{E}[T_k|x] - y_k(x))^2 f_{X|t}(x)dx +$$

$$+ \sum_{k=1}^{c} \int_{X|T} \left(\mathbb{E}[T_k^2|x] - \mathbb{E}^2[T_k|x] \right) f_{X|t}(x)dx . \tag{2.7}$$

The second term of (2.7) represents a variance of the t_k and does not depend on parameter tuning. Thus, the minimization of $\hat{R}_{MSE}$ for $n \to \infty$ implies the minimization of the first term of (2.7). In optimal conditions (to be mentioned shortly), that amounts to obtaining

$$y_k(x) = \mathbb{E}[T_k|x] . \tag{2.8}$$

This result is the version for the classification setting of the general result ($Y = \mathbb{E}[Z|X]$) previously mentioned for the regression setting. Expression (2.8) can be written out in detail as:

$$y_k(x) = \mathbb{E}[T_k|x] = \sum_{i=1}^{n} t_i \, P(T_k = t_i|x) ; \tag{2.9}$$

implying, for a 0-1 coding scheme of the t_i,

$$y_k(x) = P(T_k|x) . \tag{2.10}$$

In this case, the classifier outputs are estimates of the posterior probabilities of the input data (probability that the input pattern x belongs to class ω_k, coded as t_k). The classifier (either neural network or of other type) is then able to attain the optimal Bayes error performance.

This result for L_{SE} can in fact be generalized to other loss functions $L(t, y)$ as far as they satisfy the following three conditions [193]: a) $L(t, y) = 0$, iff $t = y$; b) $L(t, y) > 0$, if $t \neq y$; c) $L(t, y)$ is twice continuously differentiable.

In general practice, however, this radiant scenario is far from being met for the following main reasons:

1. The classifier must be able to provide a good approximation of the conditional expectations $\mathbb{E}[T_k|x]$. This may imply a more complex architecture of the classifier (e.g., more hidden neurons in the case of MLPs) than is adequate for a good generalization of its performance.
2. The training algorithm must be able to reach the minimum of $\hat{R}_{MSE}$. This is a thorny issue, since one will never know whether the training process converged to a global minimum or to a local minimum instead.
3. For simple artificial problems it may be possible to generate enough data instances so that one is sure to be near the asymptotic result (2.8), corresponding to infinite instances. One would then obtain good estimates of the posterior probabilities [185, 78]. However, in non-trivial classification problems with real-world datasets, one may be operating far away of the convergence solution.
4. Finally, the above results have been derived under the assumption of noise-free data. In normal practice, however, one may expect some amount of noise in both the data and the target values. To give an example, when a supervisor has to label instances near the decision surface separating the various classes it is not uncommon that some mistakes crawl in.

We now present two simple examples. The first one illustrates the drastic influence that a small sample size, or a tiny amount of input noise, may have in finding the $\min P_e$ solution with MSE. The second one illustrates the fact that a very small deviation of the posterior probability values may, nonetheless, provoke an important change in the $\min P_e$ value. This is a consequence of the integral nature of P_e, and should caution us as to the importance of the "posterior probability approximation" criterion.

Example 2.1. The dataset (inspired by an illustration in [41]) in this two-class example with $T = \{0, 1\}$ is shown in Fig. 2.1; corresponds to uniform distributions for both classes: the x_1-x_2 domain is $[-3, -0.05] \times [0, 0.15] \cup [-0.5, -0.05] \times [0.15, 1]$ for class 0 and reflected around the $(0, 0.5)$ point for class 1.

Let us assume a regression-like classifier implementing the thresholded linear family $\mathcal{F}_W = \{\theta(f_w(x))\}$ with $f_w(x) = w_0 + w_1 x_1 + w_2 x_2$ and $\theta(y) = h(y+0.5)$, where h is the Heaviside step. Using the MSE risk, one may apply the direct parameter estimation algorithm amounting to solving the normal equations for the linear regression problem. Once $w = (w_0, \ w_1, \ w_2)$,

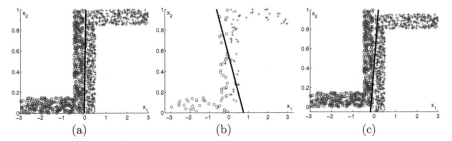

Fig. 2.1 Separating the circles from the crosses by linear regression: a) 600 instances per class; b) 120 instances per class; c) 600 instances per class with noisy x_2 of class 1. The linear discriminant solution is the solid line.

defining a planar decision surface, has been obtained, we apply the thresholding function, which in geometrical terms determines a linear *decision border* (linear discriminant) as shown in Fig. 2.1.

There are, for this classification problem, an infinity of $f_{w^*}(x)$ solutions, corresponding in terms of decision borders to any straight line inside $]-0.05, 0.05[\times[0,1]$. In Fig. 2.1a the data consists of 600 instances per class and the MMSE regression solution results indeed in one of the $\hat{P}_e = 0$ straight lines. This is the large size case; for large n (say, $n > 400$ instances per class) one obtains solutions with no misclassified instances, practically always. Figure 2.1b illustrates the small size case; the solutions may vary widely depending on the particular data sample, from close to f_{w^*} (i.e., with practically no misclassified instances) to largely deviated as in Fig. 2.1b, exhibiting a substantial number of misclassified instances. Finally, in Fig. 2.1c, the same dataset as in Fig. 2.1a was used, but with 0.05 added to component x_2 of class 1 ('crosses'); this small "noise" value was enough to provoke a substantial departure from a f_{w^*} solution, in spite of the fact that the data is still linearly separable. The error rate in Fig. 2.1c instead of zero is now above 3%. □

Example 2.2. Let us assume a univariate two-class problem (input X), with Gaussian class conditionals $f_{X|0}$ (left class) and $f_{X|1}$ (right class), with means 0 and 1 and standard deviation 0.5. The classifier task is to determine the best separating x point. Such a classifier is called a data splitter. With equal priors the posterior probabilities, $P_{T|x}$, of the classifier (see formula (1.6)) are as shown with solid line in Fig. 2.2. Note that by symmetry $P_{0|x} = 1 - P_{1|x}$ and the min P_e split point (the decision border) is 0.5.

Now suppose that due to some implementation "noise" one computed posteriors $P'_{T|x}$ with a deviation δ such that $P'_{1|x} = P_{1|x} - \delta$ for $x \in [P_{1|x}^{-1}(\delta), 0.5]$ and $P'_{1|x} = P_{1|x} + \delta$ for $x \in]0.5, 1 - P_{1|x}^{-1}(\delta)]$. Below δ, $P'_{T|x} = 0$, and above $1 - P_{1|x}^{-1}(\delta)$, $P'_{T|x} = 1$. With $P'_{0|x} = 1 - P'_{1|x}$ and $\delta = 0.01$ ($P_{1|x}^{-1}(\delta) = -0.64$) we obtain the dotted line curves shown in Fig. 2.2. The new $P'_{T|x}$ are perfect legitimate posterior probabilities and differ from $P_{T|x}$ no more than 0.01.

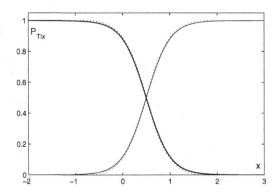

Fig. 2.2 The posterior probabilities $P_{T|x}$, (solid line) and $P'_{T|x}$ (dotted line) of Example 2.2. The differences are barely visible.

For the $P_{T|x}$ we have $\min P_e = 0.317$. By construction, the optimal split point for the $P'_{T|x}$ is also 0.5 with a variation of the $\min P_e$ value of

$$2\delta(0.5 - P_{1|x}^{-1}(\delta)) = 0.023.$$

We then obtain a deviation of the error value quite more important than the deviation of the posterior probabilities. For more complex problems, with more dimensions, one may expect sizable differences in the $\min P_e$ values between the ideal and real models, even when the PDF estimates are good approximations of the real ones.   □

Relatively to Example 2.1, one should be aware that poor MMSE solutions are not necessarily a consequence of using a direct parameter estimation algorithm. Iterative optimization algorithms may also fail to produce good approximations to the $\min P_e$ classifier, even for noise-free data. This was shown in [32] for square-error and gradient-descent applied to very simple problems.

The square-error function is a particular case of the Minkowski p-power distance family

$$L_p(t(x), y(x)) = |t(x) - y(x)|^p, \; p = 1, 2, \ldots \tag{2.11}$$

We see that $L_{SE} \equiv L_2$. The L_p family of distance-based loss functions has a major drawback for $p > 2$: it puts too much emphasis on large $t - y$ deviations, thereby rendering the risk too much sensitive to *outliers* (atypical instances), reason why no $p > 2$ distance function L_p has received much attention for machine learning applications. The square-error function itself is also sometimes blamed for its sensitivity to large $t - y$ deviations.

For $p = 1$, it is known that the regression solution to the estimation problem corresponds to finding the median of $Z|X$, instead of the mean. This could be useful for applications; however, the discontinuity of the L_1 gradient at zero raises difficulties to iterative optimization algorithms. As a matter of fact, the large popularity of the MSE risk functional stems from the existence of efficient optimization algorithms, particularly those based on the original adaptive training process known as the least-mean-square Widrow-Hoff algorithm (see e.g., [142]).

2.1.2 The Cross-Entropy Risk

The cross-entropy (CE) loss function was first proposed (although without naming it that way) in [22]; it can be derived from the maximum likelihood (ML) method applied to the estimation of the posterior probabilities $P(T_k|x)$, $k = 1, \ldots, c$, for any $x \in X$. Each component y_k of the classifier output vector, assumed as taking value in $[0, 1]$, is viewed as an estimate of the posterior probability $P(T_k|x)$; i.e., $y_k = \hat{P}(T_k|x)$.

Let us denote the $P(T_k|x)$ simply by p_k. The occurrence of a target vector t conditioned on a given input vector x, in other words, a realization of the r.v. $T|x$, is governed by the joint distribution of $(T_1|x, \ldots, T_c|x)$. For 0-1 coding the probability mass function of $T|x$ is multinomial with

$$P(T|x) = p_1^{t_1} p_2^{t_2} \ldots p_c^{t_c} . \tag{2.12}$$

Note that for $c = 2$ formula (2.12) reduces to a binomial distribution, e.g. of T_1, as

$$P(T|x) = p_1^{t_1}(1 - p_1)^{(1-t_1)} . \tag{2.13}$$

Similarly, we assign a probabilistic model to the classifier outputs, by writing

$$P(Y|x) = y_1^{t_1} y_2^{t_2} \ldots y_c^{t_c}, \text{ with } y_k = P(Y_k|x) , \tag{2.14}$$

with the assumption that the outputs satisfy the same constraints as true probabilities do, namely $\sum_k P(Y_k|x) = 1$.

We would like the $Y|x$ distribution to approximate the target distribution $T|x$. For this purpose we employ a loss function that maximizes the likelihood of $Y|x$ or, equivalently, minimizes the Kullback-Leibler (KL) divergence of $Y|x$ with respect to $T|x$ (see Appendix A).

The empirical estimate of the KL divergence for i.i.d. random variables is written in the present case as:

$$\hat{D}_{KL}(p\|y) = \frac{1}{n} \sum_{i=1}^{n} \ln \frac{P(T_i|x_i)}{P(Y_i|x_i)} = \frac{1}{n} \sum_{i=1}^{n} \ln \left(\frac{p_{i1}^{t_{i1}} \ldots p_{ic}^{t_{ic}}}{y_{i1}^{t_{i1}} \ldots y_{ic}^{t_{ic}}} \right) =$$

$$= -\frac{1}{n} \sum_{i=1}^{n} \sum_{k=1}^{c} t_{ik} \ln(y_{ik}) + \frac{1}{n} \sum_{i=1}^{n} \sum_{k=1}^{c} t_{ik} \ln(p_{ik}) . \tag{2.15}$$

Note that, since the $p_{ik} = P(T_k|x_i)$ are unknown, (2.15) cannot be used as a risk estimator. However, the p_{ik} do not depend on the classifier parameter vector w, therefore the minimization of (2.15) is equivalent to the minimization of

$$\hat{R}_{CE}(y) = -\sum_{i=1}^{n} \sum_{k=1}^{c} t_{ik} \ln(y_{ik}) \ . \tag{2.16}$$

The empirical risk (2.16) is known in the literature as the *cross-entropy* (CE) risk. This designation is, however, a misnomer. Despite the similarity between (2.16) and the cross-entropy of two discrete distributions, $-\sum_X P(x) \ln Q(x)$, with PMFs $P(x)$ and $Q(x)$, one should note that the t_{ik} are *not* probabilities (the t_i are random vectors with multinomial distribution). There is a tendency to "interpret" the t_{ik} as $P(T_k|x_i)$ and some literature is misleading in that sense. As a matter of fact, since the t_{ik} are binary-valued (in $\{0,1\}$ for the 0-1 coding we are assuming) such "interpretation" is *incorrect* (no matter which coding scheme we are using): it would amount to saying that every object is correctly classified! Briefly, the t_{ik} do *not* form a valid probability distribution. They should be interpreted as mere switches: when a particular t_{ik} is equal to 1 (meaning that x_i belongs to class ω_k), y_{ik} should be maximum and we then just minimize $-\ln(y_{ik})$, since all the remaining t_{il}, with $l \neq k$, are zero.

Although, as we have explained, the designation of (2.16) as cross-entropy is incorrect, we will keep it given its wide acceptance.

When applying the empirical $\hat{R}_{CE}$ risk, one should note that whenever the classifier outputs are continuous and differentiable, $\hat{R}_{CE}$ is also continuous and differentiable. The usual optimization algorithms can then be applied to the minimization of the empirical cross-entropy risk, namely any gradient descent algorithm.

From the above discussion it would seem appropriate to always employ a minimum cross-entropy (MCE) approach to train classifiers because when interpreting the outputs as probabilities this is the optimal solution (in a maximum likelihood sense). In fact, $\hat{R}_{CE}$ takes into account the binary characteristic of the targets. No similar interpretation exists for $\hat{R}_{MSE}$. (The ML equivalence to MSE is only valid for zero-mean and equal variance Gaussian targets.)

The derivation of $\hat{R}_{CE}$ can be found in the works of [83, 185], applying either the maximum likelihood or maximum mutual information principles, and assuming the classifier outputs are approximations of posterior probabilities. The analysis provided by [89] goes further and presents a general expression that any loss function should satisfy so that $y_k = \hat{P}(T_k|x)$. It assumes the independence of the target components t_k (a condition that is never fulfilled, since any component is the complement of all other ones) and in addition that the empirical risk is expressed as a distance functional of outputs and targets as follows:

$$\hat{R} = \sum_{i=1}^{n} \sum_{k=1}^{c} f(|t_{ik} - y_{ik}|) \ . \tag{2.17}$$

With these conditions, [89] shows that for outputs $y_k \in [0,1]$, f must asymptotically (in the limit of infinite data) satisfy

$$\frac{f'(1-y)}{f'(y)} = \frac{1-y}{y} \ , \tag{2.18}$$

implying

$$f(y) = \int y^r (1-y)^{r-1} dy \ . \tag{2.19}$$

For $r = 1$ the square-error function is obtained. For $r = 0$ one obtains $f(y) = -\ln(1 - |y|)$ leading to the CE risk (see also [26]).

Again, this favorable scenario implying that a classifier using MSE or CE would be able to attain Bayes performance provided it had a sufficiently complex architecture, is never verified in practice for several reasons (explained in the previous section), the most obvious one being that target components are not independent.

2.2 Risk Functionals Reappraised

Risk functionals are usually presented and analyzed in the literature as expectations of loss functions relative to joint distributions in the $X \times T$ space. We may, however, express risk functionals and analyze their properties relative to other spaces, functionally dependent on X and T. In fact, in order to appreciate how the various risk functionals cope with the classifier problem, it is obviously advantageous to express them in terms of the error r.v. [150,219]. We now proceed to do exactly this for the MSE and CE risk functionals. For simplicity, we will restrict the analysis to two-class problems.

2.2.1 The Error Distribution

Assuming w.l.o.g. a $\{-1,1\}$ coding of the targets, we first derive the cumulative distribution function of the error[2] r.v., $E = T - Y$, denoting by $p = P(T = 1)$ and $q = 1 - p = P(T = -1)$ the class priors, as follows:

---

[2] The "error" r.v. is indeed a deviation r.v., not the misclassification rate r.v.

$$\begin{aligned}
F_E(e) &= P(E \le e) = P((T = 1, E \le e) \vee (T = -1, E \le e)) = \\
&= P(T = 1)P(E \le e|T = 1) + P(T = -1)P(E \le e|T = -1) = \\
&= pP(1 - Y \le e|T = 1) + qP(-1 - Y \le e|T = -1) = \\
&= p(1 - F_{Y|1}(1 - e)) + q(1 - F_{Y|-1}(-1 - e)) = \\
&= 1 - pF_{Y|1}(1 - e) - qF_{Y|-1}(-1 - e) \, .
\end{aligned} \tag{2.20}$$

If the classifier output distribution is continuous, it will have a probability density function written as

$$f_Y(y) = q f_{Y|-1}(y) + p f_{Y|1}(y) \, . \tag{2.21}$$

The corresponding density of E is obtained by differentiation of $F_E(e)$:

$$f_E(e) = \frac{dF_E}{de}(e) = p f_{Y|1}(1 - e) + q f_{Y|-1}(-1 - e) \, . \tag{2.22}$$

Note that in the literature of SVMs other definitions of error are used, namely $E = |1 - TY|_+$ with $|z|_+ = \max(0, z)$; such definitions do not provide, in general, continuous error distributions even when the output class-conditional $f_{Y|t}$ are continuous.

Let us now consider two-class classifiers such that for the $\{-1, 1\}$ target coding, the classifier output, Y, is restricted to the $[-1, 1]$ interval. As a matter of fact, we will see at a later section (2.3.1) why such an *interval codomain* restriction (imposed e.g. by sigmoid functions at neural network outputs) turns out to be advantageous. In this case $E = T - Y$ takes value in $[-2, 2]$. If the class-conditional PDFs of Y have no discontinuities at support ends an important constraint on f_E is imposed: $f_E(0) = 0$ since $\lim_{\epsilon \to 0} f_{Y|-1}(-1 + \epsilon) = \lim_{\epsilon \to 0} f_{Y|1}(1 + \epsilon) = 0$, as illustrated in Fig. 2.3. Also note that $e \in [0, 2]$ for $f_{Y|1}$ and $e \in [-2, 0]$ for $f_{Y|-1}$.

Equivalent expressions to (2.20) and (2.22) are obtained for other target coding schemes.

Note that in order to achieve $P_e = 0$ one only needs to guarantee a null area in a subset of the codomain. In the case of Fig. 2.3, let us assume a classifier thresholding function $\theta(y) = 2h(y) - 1$, where h is the Heaviside function (equivalent to classifying as 1 if $y \ge 0$ and -1 otherwise). Denoting by a, b, c, and d the $f_E(e)$ areas respectively in the intervals $[-1, 0]$, $[-2, -1]$, $[1, 2]$, and $[0, 1]$ as in Fig. 2.3 (right), one obtains the decision Table 2.1. Clearly $P_e = 0$ implies $b + c = 0$.

2.2.2   MSE and CE Risks in Terms of the Error

Note that the risk functional in (2.1) is expressed in terms of $f_{X|t}(x)$, whereas in formula (2.22) we expressed $f_E(e)$ in terms of the class-conditional PDFs of Y. In order to express the risk functionals in terms of $f_E(e)$ we first express

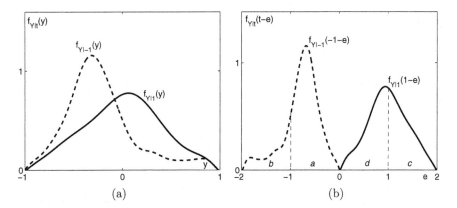

Fig. 2.3 Illustration of the transformation $E = T - Y$ for a two-class classification problem, emphasizing the fact that $f_E(0) = 0$ for continuous class-conditionals.

Table 2.1 Decision table corresponding to Fig. 2.3b.

		Classifier output	
		-1	1
True	-1	a	b
class	1	c	d

them in terms of $f_Y(y)$ using the following theorem.

Theorem 2.1. *Suppose X_1, X_2, $\ldots$, X_k are continuous random variables and $Y = g(X_1, X_2, \ldots, X_k)$ for some function g. Suppose also that*

$$\int_{X_1} \cdots \int_{X_k} |g(x_1, \ldots, x_k)| f_{X_1, \ldots, X_k}(x_1, \ldots, x_k) dx_1 \ldots dx_k < \infty . \quad (2.23)$$

Then

$$\mathbb{E}[Y] = \int_{X_1} \cdots \int_{X_k} g(x_1, \ldots, x_k) f_{X_1, \ldots, X_k}(x_1, \ldots, x_k) dx_1 \ldots dx_k . \quad (2.24)$$

$\square$

For a proof of this theorem see, for instance, [61, 87]. Theorem 2.1 allows us to compute an expected value of an r.v. Y either directly according to the definition (i.e., in terms of $f_Y(\cdot)$), or whenever Y is given by a certain function of other r.v.'s as in (2.24). The class-conditional expected value expressed in (2.3) — where X may be a multidimensional r.v. — is then simply $E_{Y|t}[L(t, Y)]$. Therefore, assuming Y restricted to $[-1, 1]$, we may write

$$R_L(Y) = \sum_{t \in T} P(t)\mathbb{E}_{Y|t}[L(t,Y)] = \sum_{t \in \{-1,1\}} P(t) \int_{-1}^{1} L(t,y)f_{Y|t}(y)dy \ , \quad (2.25)$$

if the absolute integrability condition (2.23) for $L(t,y)$ is satisfied.

Applying again Theorem 2.1 to $E = T - Y$, the risk functional (2.25) is finally expressed in terms of the error variable as

$$R_L(E) = \sum_{t \in \{-1,1\}} P(t) \int_{t-1}^{t+1} L(t,e)f_{E|t}(e)de \ . \quad (2.26)$$

For MSE, $L_{SE}(t,e) = (t-y)^2 = e^2$ depends only on e (or, in more detail, $e_w = t - y_w$). We then have $R_{MSE}(E) = \mathbb{E}_{T,E}[E^2]$, the second order moment of the error, which is empirically estimated as in (2.5) and can be rewritten as

$$\hat{R}_{MSE}(Y) = \frac{1}{n} \left(\sum_{t_i=-1} (t_i - y_i)^2 + \sum_{t_i=1} (t_i - y_i)^2 \right) \ . \quad (2.27)$$

Let us now consider the cross-entropy risk whose empirical estimate is given by formula (2.16). For a two-class problem and the $\{0,1\}$-coding scheme one obtains the following popularized expression, when the classifier has a single output:

$$\hat{R}_{CE}(Y) = -\sum_{t_i=0} (1-t_i)\ln(1-y_i) - \sum_{t_i=1} t_i\ln(y_i) \ . \quad (2.28)$$

The $\{-1,1\}$-coding implies a $y \to (y+1)/2$ transformation; formula (2.28) is then rewritten as

$$\hat{R}_{CE}(Y) = -\sum_{t_i=-1} \ln\left(\frac{1-y_i}{2}\right) - \sum_{t_i=1} \ln\left(\frac{1+y_i}{2}\right) \ . \quad (2.29)$$

When multiplied by n, $\hat{R}_{CE}(Y)$ can be viewed as the empirical estimate of the following (theoretical) risk functional:

$$R_{CE}(Y) = -P(-1) \int_{-1}^{1} \ln(1-y)f_{Y|-1}(y)dy-$$
$$- P(1) \int_{-1}^{1} \ln(1+y)f_{Y|1}(y)dy + \ln(2) \ . \quad (2.30)$$

Applying the same variable transformation as we did before, the CE risk functional is finally expressed in terms of the error variable as

$$R_{CE}(E) = \sum_{t \in \{-1,1\}} P(t) \int_{t-1}^{t+1} \ln\left(\frac{1}{2-te}\right) f_{E|t}(e)de + \ln(2) \ . \quad (2.31)$$

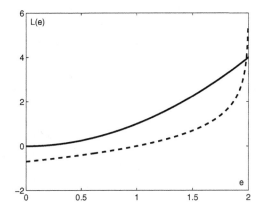

Fig. 2.4 L_{MSE} (solid line) and L_{CE} (for $t = 1$) (dashed line) as functions of e.

Thus, for cross-entropy we have a logarithmic loss function, $L_{CE}(t, e) = -\ln(2 - te)$. Figure 2.4 shows (for $t = 1$) the distance functions $L_{MSE}(e) = e^2$ and $L_{CE}(t, e) = -\ln(2 - te)$. A classifier minimizing the MSE risk functional is minimizing the second-order moment of the errors, favoring input-output mappings with low error spread (low variance) and deviation from zero. A classifier minimizing the CE risk functional is minimizing an average logarithmic distance of the error from its worst value (respectively, 2 for $t = 1$ and -2 for $t = -1$), as shown in Fig. 2.4. As a consequence of the logarithmic behavior $L_{CE}(t, e)$ tends to focus mainly on large errors. Note that one may have in some cases to restrict Y to the open interval $]-1, 1[$ in order to satisfy the integrability condition of (2.31).

Example 2.3. Let us consider a two-class problem with target set $T = \{0, 1\}$, $P(0) = P(1) = 1/2$ and a classifier codomain restricted to $[0, 1]$ according to the following family of uniform PDFs:

$$f_Y(y) = \begin{cases} u(y; 0, d) & \text{if } T = 0 \\ u(y; 1 - d, 1) & \text{if } T = 1 \end{cases} . \tag{2.32}$$

Note that according to (2.22) $f_E(e)$ is distributed as $\frac{1}{2}u(e; 0, d) + \frac{1}{2}u(e; -d, 0) = u(e; -d, d)$. Therefore, the MSE risk is simply the variance of this distribution:

$$R_{MSE}(d) = \frac{d^2}{3}. \tag{2.33}$$

We now compute the cross-entropy risk, which is in this case easier to derive from $f_{Y|t}$. First note that (2.28) is n times the empirical estimate of

$$R_{CE}(d) = -P(0)\mathbb{E}[\ln(1 - Y)|T = 0] - P(1)\mathbb{E}[\ln(Y)|T = 1] . \tag{2.34}$$

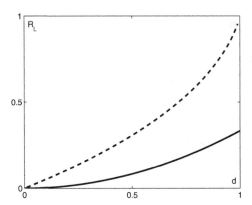

Fig. 2.5 The dependence of R_{MSE} (solid line) and R_{CE} (dashed line) on d (Example 2.3).

Given the dataset symmetry, we have:

$$R_{CE}(d) = -2\left[\frac{1}{2}\mathbb{E}[\ln(Y)|T=1]\right] = \frac{1}{d}((1-d)\ln(1-d)+d) . \qquad (2.35)$$

Figure 2.5 shows the dependence of R_{MSE} and R_{CE} on d. A large d corresponds to a large overlap of the distributions. Note the penalization of larger errors in accordance to Fig. 2.4, and the special cases $d = 0$ (maximum class separation with Dirac-δ distributions) and $d = 1$ (no class separation). In this example the integrability condition for R_{CE} is satisfied for the whole codomain.   □

2.3 The Error-Entropy Risk

2.3.1 EE Risks

The risk functional constituting the keystone of the present book is simply the entropy of the error r.v., E, the error entropy, EE. Using the Shannon differential entropy, the Shannon error-entropy (SEE) risk is written as

$$R_{SEE}(E) \equiv H_S(E) = -\int_E f(e)\ln f(e)de . \qquad (2.36)$$

Other formulations of the differential entropy can, in principle, be used. One interesting formulation is the Rényi entropy [183] of order α:

$$H_{R_\alpha}(E) = \frac{1}{1-\alpha} \ln \int_E f^\alpha(e)de, \ \alpha \geq 0, \ \alpha \neq 1 \ . \tag{2.37}$$

For $\alpha \to 1$ one obtains the Shannon entropy. We will find it useful to use Rényi's quadratic entropy ($\alpha = 2$), expressing the risk as

$$R_{R_2EE}(E) \equiv H_{R_2}(E) = -\ln \int_E f^2(e)de \ . \tag{2.38}$$

Until now we have considered continuous distributions of the errors. In some problems, however, one has to deal with a discrete error r.v.; one then uses the discrete versions of the entropies (historically, simply called entropies):

$$H_S(E) = -\sum_{i=1}^m P(e_i) \ln P(e_i) \ , \tag{2.39}$$

$$H_{R_2}(E) = -\ln \sum_{i=1}^m P^2(e_i) \ , \tag{2.40}$$

where $P(e) \equiv P_E(e)$ is the error PMF.

We shall later see how to estimate $H_S(E)$ and $H_{R_2}(E)$ for the continuous error case. We shall also see that when applying $H_S(E)$ formula (2.36) to data classification a crude and simple estimation of $f(e)$ is all that is required; moreover, when applying $H_{R_2}(E)$ the estimation of $f(e)$ is even short-circuited. Thus, in both cases we get rid of having to accurately estimate a PDF, a problem that has traditionally been considered as being more difficult, in general, than that of having to design an accurate classifier [41, 227].

Note that indeed one may interpret the above entropies as risk functionals. $R_{SEE}(E)$ is the expectation of the loss function $L_{SEE}(e) = -\ln f(e)$. For Rényi's quadratic error-entropy, instead of minimizing $R_{R_2EE}(E)$ we will see later that it turns out to be more convenient to maximize $V_{R_2}(E) = \exp(-R_{R_2EE}(E))$, the so-called *information potential* [175]. In this case, instead of a loss function we may speak of a *gain* function: $V_{R_2}(E)$ is the expectation of the gain function $f(e)$. One can also, of course, consider $-V_{R_2}(E)$ as the risk functional expressed in terms of a loss $-f(e)$.

As an initial motivation to use entropic risk functionals, let us recall that entropy provides a measure of how concentrated a distribution is. For discrete distributions its minimum value (zero) corresponds to a discrete Dirac-δ PMF. For continuous distributions the minimum value (minus infinite for H_S and zero for H_{R_2}) corresponds to a PDF represented by a sequence of continuous Dirac-δ functions, a Dirac-δ comb. Let us consider the 1-of-c coding scheme with $T \in \{a, b\}$. For an interval-codomain regression-like classifier, implementing an $X \to T$ mapping, the class-conditional densities of any output $Y_k \in [a, b]$ are Dirac-δ combs iff they are non-null only at $\{a, b\}$. Let us assume equal priors and b as the label value of ω_k and a of its complement. As an example of a Dirac-δ comb for the error PDF, we could have $Y_k = b$, i.e., assigning all input instances to ω_k; in this case, $f_E(e) = \frac{1}{2}\delta(e-(a-b))+\frac{1}{2}\delta(e)$. A single Dirac-$\delta$ for the error PDF can only be obtained, and at the origin, when all errors are zero. This is an ideal $\min P_e = 0$ situation, demanding a classifier function family sufficiently rich to allow such an error PDF, and in case of iterative training requiring an algorithm which does indeed guarantee the convergence of the error PDF towards the single Dirac-δ at the origin ($E = 0$).

2.3.2 EE Risks and Information Theoretic Learning

The use of the entropy concept in data classification is not new. It has been, namely, used as a measure of efficient dataset partition in clustering applications and as a node splitting criteria in decision trees, to be described at a later section [237, 33]. It has also been proposed for classifier selection [237, 238]. The novelty lies here in its use as a risk functional, and the analysis of how it performs both theoretically and experimentally, when applied to classifier design.

The MEE approach to classifier design fits into the area of Information Theoretic Learning (ITL), a research area enjoying a growing interest and promising important advances in several applications. The introduction of information theoretic measures in learning machines can be traced back at least to [237, 238] and to [140]; this last author introduced the maximization of mutual information between input and output of a neural network (the infomax principle) as an unsupervised method that can be applied, for instance, to feature extraction (see also [49]). A real blossom of ITL in the areas of learning systems and of signal processing came in more recent years, when J.C. Príncipe and co-workers built a large amount of theoretical and experimental results of entropic criteria applied to both areas. In particular, they proposed the minimization of Rényi's quadratic entropy of data errors for solving regression problems [66], time series prediction [67], feature extraction [84,117], and blind source separation [99,65,67] using adaptive systems. These and more recent advances on ITL issues and applications can be found in the monograph [174].

The rationale behind the application of MEE to learning systems performing a regression task is as follows: the MEE approach implies a reduction of the expected information contained in the error, leading to the maximization of the mutual information between the desired target and the system output [66,67]. This means that the system is learning the target variable.

Information Theory also provides an interesting framework for analyzing and interpreting risk functionals, as shown in the work of [37]. These authors demonstrated the following

Theorem 2.2. *Given any loss function $L(e)$, which satisfies $\lim_{|e| \to +\infty} L(e)$*
$= +\infty$, one has $R_L(E) \propto \{H_S(E) + D_{KL}(f_E(e) \| q_L(e))\}$, where $q_L(e)$ is a
PDF related to $L(e)$ by $q_L(e) = \exp(-\gamma_0 - \gamma_1 L(e))$.   □

The proof of the theorem and the existence of $q_L(e)$ is demonstrated in the
cited work, which also provides the means of computing the γ_0 and γ_1 constants. Note that all loss functions we have seen so far, with the exception
of the one corresponding to Rényi's quadratic entropy, satisfy the condition
$\lim_{|e| \to +\infty} L(e) = +\infty$. Theorem 2.2 also provides an interesting bound on
$R_L(E)$. Since the Kullback-Leibler divergence is always non-negative, we have

$$H_S(E) + D_{KL}(f_E(e) \| q_L(e)) \geq H_S(E) , \qquad (2.41)$$

with equality iff $f_E(e) = q_L(e)$. Therefore, minimizing any risk functional
$R_L(E)$, with $L(e)$ satisfying the above condition, is equivalent to minimizing
an upper bound of the error entropy $H_S(E)$. Moreover, Theorem 2.2 allows us
to interpret the minimization of any risk functional $R_L(E)$ as being driven by
two "forces": one, $D_{KL}(f_E(e) \| q_L(e))$, that attempts to shape the error PDF
in a way that reflects the loss function itself; the other, $H_S(E)$, providing the
decrease of the dispersion of the error, its uncertainty.

There is an abundant literature on information theoretic topics. For the
reader unfamiliar with this area an overview on definitions and properties
of entropies (Shannon, generalized Rényi, and others) can be found in the
following works: [131, 183, 168, 48, 184, 164, 96, 62]. Appendix B presents a
short survey of properties that are particularly important throughout the
book.

2.3.3 MEE Is Harder for Classification than for Regression

An important result concerning the minimization of error entropy was shown
in [67] for a machine solving a regression task, approximating its output y to
some desired continuous function $d(x)$. The authors showed that the MEE
approach corresponds to the minimum of the Kullback-Leibler divergence of
$f_{X,Y}$ (the joint PDF when the output is $y_w(x)$) with respect to $d_{X,Y}$ (the
joint PDF when the output is the desired $d(x)$). Concretely, they showed that

$$\min H_S(E) \Rightarrow \min D_{KL}(f_{X,Y} \| d_{X,Y}) = \int_X \int_Y f_{X,Y}(x, y) \ln \frac{f_{X,Y}(x, y)}{d_{X,Y}(x, y)} dx dy .$$
$$(2.42)$$

Their demonstration was, in fact, presented for the generalized family of Rényi
entropies, which includes the Shannon entropy $H_S(E)$ as a special asymptotic
case. Moreover, although not explicit in their demonstration, the above result
is only valid if the above integrals exist, which among other things require

$d_{X,Y}(x, y) \neq 0$ in the (X, Y) domain. The "probability density matching" expressed by (2.42) is of course important for regression applications.

For data-classification the application of the MEE approach raises three difficulties [219]. We restrict our discussion to two-class problems and use therefore (2.22) as the error PDF. We also assume the interval codomain restriction implying that each class-conditional density $f_{Y|t}(t - e)$ lies in separate $[t - 1, t + 1]$ intervals. As a consequence the differential Shannon's entropy of the error $H_S(E)$ can be decomposed as

$$H_S(E) = pH_{S|1}(E) + qH_{S|-1}(E) + H_S(T) , \qquad (2.43)$$

where $H_{S|t}$ is the Shannon's entropy of the error for class ω_t and $H_S(T) = \sum_{t \in T} P(t) \ln P(t)$ is the Shannon's entropy of the priors $(P(1) \equiv p, P(-1) \equiv q)$. Rényi's quadratic entropy also satisfies a similar additive property when exponentially scaled (see Appendix C for both derivations). Let us recall that class conditional distributions and entropies depend on the classifier parameter w, although we have been omitting this dependency for the sake of simpler notation.

The first difficulty when applying MEE to data classification has to do with expression (2.43). Since $H_S(T)$ is a constant, $\min H_S(E)$ implies $\min[pH_{S|1}(E) + qH_{S|-1}(E)]$. Thus, in general, one can say nothing about the minimum (location and value) of H_S since it will depend on the particular shapes of $H_{S|t}$ as functions of w, and the particular value of p. For instance, an arbitrarily low value of H_S can be achieved if one of the $f_{Y|t}$ is arbitrarily close to a Dirac-δ distribution even if the other has the largest possible entropy (i.e., is a Gaussian distribution, under specified variance).

The second difficulty has to do with the fact that for the regression setting one may write $f_E(e) = f_{Y|x}(d - e)$ as in [67], since there is only one distribution of y values and d can be seen as the average of the y values. However, for the classification setting one has to write $f_{E|t}(e) = f_{Y|t,X}(d - e, x)$. That is, one has to study what happens to each class-conditional distribution individually and, therefore, to individually study the KL divergence relative to each class conditional distribution, that is:

$$D_{KL}(f_{X,Y|t} \| d_{X,Y|t}) = \int_X \int_Y f_{X,Y|t}(x, y) \ln \frac{f_{X,Y|t}(x, y)}{d_{X,Y|t}(x, y)} dx dy , \qquad (2.44)$$

where $d_{X,Y|t}(x, y)$ is the desired joint probability density function for class ω_t.

Finally, the third difficulty arises from the fact that the KL divergence does not exist whenever $d_{X,Y|t}(x, y)$ has zeros in the domains of X and Y. This problem, which may or may not be present in the regression setting, is almost always present in the classification setting, since the desired input-output probability density functions are usually continuous functions with zeros in their domains; namely, one may desire the $d_{X,Y|t}(x, y)$ to be Dirac-δ functions.

Even if we relax the conditions on the desired probability density functions, for instance, by choosing functions with no zeros on the Y support but conveniently close to Dirac-δ functions, we may not yet reach the MEE condition for classification because of (2.43): attaining the KL minimum for one class conditional distribution, says nothing about the other class conditional distribution and about H_S.

2.3.4 The Quest for Minimum Entropy

We have presented some important properties of R_{MSE} and R_{CE} in Sect. 2.2. We now discuss the properties of the R_{SEE} risk functional for classification problems. Restricting ourselves to the two-class setting with codomain restriction and $T = \{-1, 1\}$, we rewrite (2.43) as

$$H_S(E) = \sum_{t \in \{-1,1\}} P(t) \int_{t-1}^{t+1} \ln\left(\frac{1}{f_{E|t}(e)}\right) f_{E|t}(e) de + H_S(T) . \qquad (2.45)$$

We see that $L_{EE_t}(e) = -\ln f_{E|t}(e)$ are here the loss functions for the two classes. The difference relative to L_{SE} and L_{CE} (and other conventional, distance-like, loss functions) is that in this case the loss functions are expressed in terms of the unknown $f_{E|t}(e)$. Furthermore, in adaptive training of a classifier $f_{E|t}(e)$ will change in unforeseeable ways. The same can be said of Rényi's quadratic entropy, with gain function $f_{E|t}(e)$. Therefore, the properties of the entropy risk functionals have to be analyzed not in terms of loss functions but of the entropies themselves.

Although pattern recognition is a quest for minimum entropy [237], the topic of entropy-minimizing distributions has only occasionally been studied, namely in relation to finding optimal locations of PDFs in a mixture [115, 38] and applying the MinMax information measure to discrete distributions [251]. Whereas entropy-maximizing distributions obeying given constraints are well known, minimum entropy distributions on the real line are often difficult to establish [125]. The only basic known result is that the minimum entropy of unconstrained continuous densities corresponds to Dirac-δ combs (sequences of Dirac-δ functions, including the single Dirac-δ function); for discrete distributions the minimum entropy is zero and corresponds to a single discrete Dirac-δ function.

Entropy magnitude is often thought to be associated with the magnitude of the PDF tails, in the sense that larger tails imply larger entropy. (A PDF $f(\cdot)$ has larger right tail than PDF $g(\cdot)$ for positive x if $\exists x_0, \forall x > x_0, f(x) > g(x)$; similarly, for left tail.) However, this presumption fails even in simple cases of constrained densities: the unit-variance Gaussian PDF, $g(x; 0, 1)$, has smaller tails than the unit-variance bilateral-exponential PDF, $e(x; \sqrt{2}) = \exp(-\sqrt{2}|x|)/\sqrt{2}$; however, the former has larger Shannon entropy, $\sqrt{2\pi e} =$

2.84, than the latter, $1 + \ln(\sqrt{2}) = 2.41$. In spite of these difficulties, one can still present three properties of H_S that are useful in guiding the search of experimental settings and in the interpretation of experimental results:

1. $H_S(E)$ is invariant to translations and partitions of $f_E(e)$. This property, a direct consequence of the well-known entropy invariance to translations, and of a result on density partitions presented in Appendix C, is illustrated in Fig. 2.6. This is a property that may lead MEE to perform worse than MMSE or MCE in cases where an equal probability of error may correspond to distinct configurations.
2. For a large number of PDF families H_S increases with the variance. This property (illustrated in Fig. 2.6) is analyzed in Appendix D and is associated with the common idea that *within the same density family*, "longer tails" (resulting in larger variance) imply larger entropy. Although there are exceptions to this rule, one can quite safely use it in the case of $f_E(e)$ densities. As a consequence, one can say that MEE favors the "order" of the errors (large H_S meaning "disorderly" errors, i.e., not concentrated at one location).
3. Whenever $f_E(e)$ has two components of equal functional form and equal priors, for a large number of $f_E(e)$ families $H_S(E)$ will decrease when the smaller (or equal) variance component decreases while the variance of the other component increases by the same amount (keeping the functional form unchanged). This property (illustrated in Fig. 2.6) is a consequence of the fact that entropy is an up-saturating function of the variance for a large number of PDF families (see Appendix D where the definition of up-saturating function is also presented); therefore, when the larger variance component dilates the corresponding increase in entropy is outweighed by the decrease in entropy of the other component. As a consequence of this property MEE is more tolerant than MMSE or MCE to tenuous tails or outlier errors, as exemplified in the following section.

Example 2.4. We now compute the Shannon entropy of error for the dataset of Example 2.3. Since $f_E(e) = u(e; -d, d)$ and the Shannon entropy of $u(x; a, b)$ is $\ln(b - a)$, we have

$$H_S(E) = \ln(2d) = \frac{1}{2} \ln(12\sigma^2) \ .$$

The entropy $H_S(E)$ is an up-saturating function (see definition in the Appendix D) of the variance σ^2. Moreover, $H_S(E)$ penalizes infinitely less the maximum separation of the classes ($f_E(e) = \delta(e)$ for $d = 0$) relatively to the total overlap ($f_E(e) = u(e; -1, 1)$). □

The error PDF family of Examples 2.3 and 2.4, parameterized by d, was such that both MSE and SEE risks attained the minimum value for $d = 0$. Both methods agreed on which PDF was most concentrated. One may

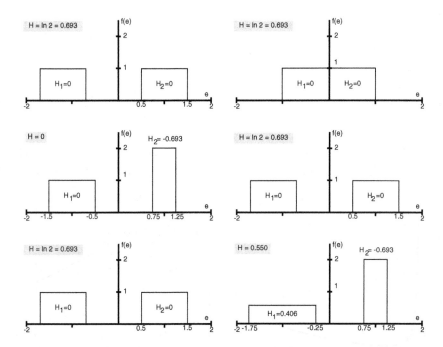

Fig. 2.6 $H_S(E)$ properties for $p = q = 1/2$. Top row (Property 1): Entropy is invariant to partitions and translations. Middle row (Property 2): For the same PDF family, an increase in variance implies an increase in entropy (see text). Bottom row (Property 3): The variance of the right component decreased by $0.25/12$ while the other increased by the same amount; however, the decrease in entropy of the right component more than compensates for the increase in entropy of the left component.

wonder whether such an agreement is universal; stating in a different way, is it possible to present examples of PDF families for which MSE and MEE do not agree on which family member is most concentrated? The answer is affirmative as we show in the following example [212].

Example 2.5. Let us consider the following family of continuous PDFs, with one parameter, $\alpha > 0$:

$$f(x; \alpha) = \frac{1}{4}[tr(x; 0, \alpha) + tr(x; -\alpha, 0) + tr(x; 0, 1/\alpha) + tr(x; -1/\alpha, 0)] , \quad (2.46)$$

where $tr(x; a, b)$ is the symmetrical triangular distribution in $[a, b]$, defined for $a \geq 0$ as

$$tr(x; a, b) = \begin{cases} \frac{4(x-a)}{(b-a)^2} & a \leq x \leq (a+b)/2 \\ \frac{4(b-x)}{(b-a)^2} & (a+b)/2 < x \leq b \end{cases} . \quad (2.47)$$

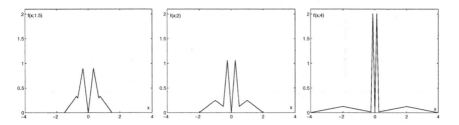

Fig. 2.7 Three members of the $f(x;\alpha)$ family of Example 2.5, with $\alpha = 1.5, 2$, and 4, from left to right.

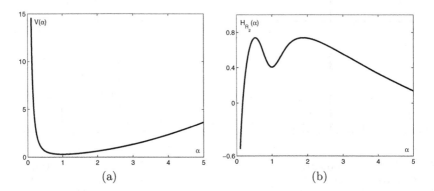

Fig. 2.8 MSE (a) and R_2EE (b) for the $f(x;\alpha)$ family of Example 2.5, as functions of α.

Figure 2.7 shows three members of the $f(x;\alpha)$ family. For $\alpha \to 0$ or $\alpha \to +\infty$ one obtains PDFs with components that are progressively more symmetric around the origin, whereas other components correspond to progressively longer tails.

Figure 2.8 shows the variance and Rényi's quadratic entropy for the $f(x;\alpha)$ family, plotted as functions of α. The variance (the same as the MSE since each family member is symmetric) has a minimum at $\alpha = 1$. If the $f(x;\alpha)$ family represented an error PDF family, $f(e;\alpha)$, one wouldn't certainly be satisfied with a convergence to the $f(e;1)$ PDF. One would surely be happier with a convergence towards $f(e;0)$ or $f(e;+\infty)$, corresponding to entropy minima.

□

2.3.5 Risk Functionals and Probability of Error

Traditionally, the role played by different risk functionals has been somewhat overlooked. There has been a persistent belief that the choice of loss function is more a computational issue than an influencing factor in system performance [41, 227]. This way of thinking has been shown in [189] to be incorrect for the unbounded codomain loss functions used by support vector machines. The authors have shown that the choice of loss function influences the convergence rate of the empirical risk towards the true risk, as well as the upper bound on $R_L(Y_{w+}) - \hat{R}_L(\hat{Y}_{w+})$, where $\hat{Y}_{w+}$ is the minimizer of the empirical risk $\hat{R}_L$.

Our interest in what regards classification systems is the probability of error. We first note that minimization of the risk does not imply minimization of the probability of error [89, 160] as we illustrate in the following example.

Example 2.6. Let us assume a two-class problem and a classifier with two outputs, one for each class: y_0, y_1. Assuming $T = \{0, 1\}$ and Y_0, Y_1 taking value in $[0, 1]$, the squared error loss for class ω_0 instances ($t_0 = 1$, $t_1 = 0$) can be written as

$$L_{MSE}([t_0\ t_1], [y_0\ y_1]) = (t_0 - y_0)^2 + (t_1 - y_1)^2 = (1 - y_0)^2 + y_1^2 . \quad (2.48)$$

The equal-L_{MSE} contours for class ω_0 instances are shown in Fig. 2.9. Let us suppose that the classifier labels the objects in such a way that an object x is assigned to class ω_0 if $y_0(x) > y_1(x)$ and to class ω_1 otherwise. Now consider two ω_0 instances x_1 and x_2 with outputs as shown in the Fig. 2.9. We see that the correctly classified x_2 ($y_0(x_2) > y_1(x_2)$) has higher L_{MSE} than the misclassified x_1. For this reason, L_{MSE} is said to be non-monotonic in the continuum connecting the best classified case ($y_0 = 1$, $y_1 = 0$) and worst classified case ($y_0 = 0$, $y_1 = 1$) [89, 160]. A similar result can be obtained for L_{CE}. This means that it is possible, at least in theory, to train a classifier that minimizes L_{MSE} and L_{CE} without minimizing the number of misclassifications [32]. □

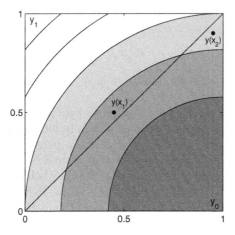

Fig. 2.9 Contours of equal-L_{MSE} for ω_0 instances. Darker tones correspond to smaller values.

Do risk functionals behave similarly with respect to the classifier problem, i.e., to the min P_e issue? The answer to this question is not easy even when we restrict the classes of (X, T) distributions and the classifier families under consideration. On one hand, none of the previously discussed risk functionals provides, in general, the min P_e solution (although they can achieve that in particular cases); on the other hand, there is no theoretical evidence precluding the existence of a risk functional that would always provide the min P_e solution.

We have seen in Sects. 2.1.1 and 2.1.2 that both MMSE and MCE although theoretically able to produce estimates of posterior probabilities (and therefore of attaining the optimal Bayes error, $P_e(Z_{Bayes})$), provided some restrictive conditions are satisfied, are unable to achieve that in practice mainly because at least some of the conditions — such as arbitrarily complex classifier architecture or independence of the target components — are unrealistically restrictive. But what really interests us is not the attainment of $P_e(Z_{Bayes})$, but of the minimum probability of error for some classifier function family, Z_W. This issue, as far as we know, has never been studied for MMSE and MCE. We will present later in the book some results on this issue for MEE.

Is it possible to conceive data classification problems where MCE and MEE perform better than MMSE? And where MEE outperforms both MCE and MMSE? The answer to these questions is affirmative, as we shall now show with a simple example of a family of data classification problems, where for an infinite subset of the family MEE provides the correct solution, whereas MMSE and MCE do not [150, 219].

Example 2.7. Let us consider a family of two-class datasets in bivariate space $\mathbb{R}^2$, target space $T = \{-1, 1\}$. We denote the input vectors by $x = [x_1 \ x_2]^T$, and assume the following marginal and independent PDFs:

$$f_1(x_1) = \frac{1}{2}[u(x_1; a, 1) + u(x_1; b, a)], \ f_{-1}(x_1) = f_1(-x_1) \ , \tag{2.49}$$

$$f_t(x_2) = u\left(x_2; -\frac{c}{2}, \frac{c}{2}\right) \ , \tag{2.50}$$

where $u(x; a, b)$ is the uniform density in $[a, b]$. We further assume $a \in [0, 1[$, $b < a$, $c > 0$ and $P(1) = P(-1) = 1/2$.

Figure 2.10 shows a sample of 500 instances per class, random and independently drawn from the above distribution family for a particular parameter choice. We see, that for suitable choices of the parameters we can obtain distributions combining a high concentration around $\{-1, 1\}$ with long tails.

Let us assume a classifier implementing the thresholded linear family $\vartheta = \{h(\varphi(\mathbf{x})) = h(\mathbf{w}^T\mathbf{x}); \ \mathbf{w} \in \mathbb{R}^2\}$, where $h(\cdot)$ is the Heaviside function. The decision surfaces are straight lines passing through the origin ($\varphi(\mathbf{x}) > 0$ assigns $\mathbf{x}$ to ω_1; otherwise to ω_{-1}). The classifier problem consists of selecting the straight line providing the min P_e solution.

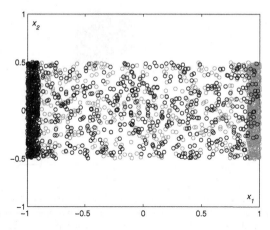

Fig. 2.10 A 500 instances per class sample distributed according to expressions (2.49) and (2.50), for $a = 0.9$, $b = -1$, $c = 1$ (class ω_1 in gray).

The solutions can be simply described by the angle α such that $x_2 = x_1 \tan \alpha$. For every α the error density can be obtained by first deriving the density of the classifier output. The theoretical Shannon EE and MSE (variance, V, since $f_E(e)$ is symmetric around the origin) are easily computed for two configurations:

- Configuration with $\alpha = -\pi/2$, $\varphi(\mathbf{x}) = -x_1$ (the min P_e solution):

$$H_S = \frac{1}{2} \ln(1 - a) + \frac{1}{2} \ln(a - b) - \ln \frac{1}{4} \; ; \qquad (2.51)$$

$$V = \frac{(1 - a)^2}{6} + \frac{(a - b)^2}{24} + \frac{(2 - a - b)^2}{8} \; . \qquad (2.52)$$

- Configuration with $\alpha = 0$, $\varphi(\mathbf{x}) = x_2$:

$$H_S = \ln c - \ln \frac{1}{2} \; ; \qquad (2.53)$$

$$V = \frac{c^2}{12} + 1 \; . \qquad (2.54)$$

For this family of datasets, MEE and MMSE do not always pick from ϑ the correct solution (the vertical line at $\alpha = -\pi/2$). Let us first restrict ourselves to the two configurations above (the classifier must either select the $\alpha = -\pi/2$ or the $\alpha = 0$ straight line). When a method picks the right solution let us call that a "win", otherwise let us call that a "loss".

Figure 2.11 shows subsets of the (a, b) space for two values of c (using the above formulas). We see that for both c values there are subsets in the (a, b) space where MEE wins and MMSE loses. Such subsets can be found for every c.

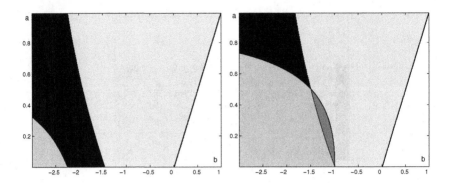

Fig. 2.11 Wins and losses of MEE and MMSE for two values of c (left: $c = 3$; right: $c = 2$) and a classifier selecting a solution out of two possible ones (see text). The tone code is as follows: ▨ - MEE and MMSE both win; ▨ - MEE and MMSE both loose; ▨ - MMSE wins and MEE looses; ■ - MEE wins and MMSE looses.

Moreover, for $c \gtrsim 2.2$ there are no subsets where MMSE wins and MEE loses.

Similar conclusions are reached when the whole interval $[-\pi/2, 0]$ is considered for α. Figure 2.12a shows how H_S and V vary with α when $a = 0.95$, $b = -1.7$, and $c = 0.9$. MEE picks the correct solution, with $\min P_e = 0.321$, whereas MMSE wrongly selects $\alpha = -0.377$ with $P_e = 0.355$. The curves were obtained by numerical simulation using a large number of instances (4000 per class) in order to obtain a very close approximation to the theoretical values of the above formulas. Although not shown in Fig. 2.12a, MCE also picks the correct solution for these parameter values. For sufficiently large tails MCE also makes a wrong decision and MEE does not. For instance, keeping the same values for a and c ($a = 0.95, c = 0.9$), the cross-entropy curve for $b = -2.4$ is shown in Fig. 2.12b. MCE selects $\alpha = -0.346$ with $P_e = 0.411$(whereas $\min P_e = 0.362$).

Essentially the same conclusions are obtained if instead of the theoretical MEE, MSE, and CE risks, we use their empirical estimates.   □

Note that this example constitutes a good illustration of entropy property 3 (presented in Sect. 2.3.3), explaining the reduced sensitivity of MEE to tenuous tails.

We then arrive to the conclusion that none of the studied risk functionals will perform in the optimal $\min P_e$ sense for all possible classes of problems. There is also no evidence that any other risk functional we could think of would always perform optimally in the $\min P_e$ sense. It then seems advisable to use classifier learning algorithms employing several types of risk functionals with different behaviors. As an alternative one could also envisage a meta-parametrized risk functional emulating the behavior of a whole set of risk functionals; such an alternative is presented in Chap. 5.

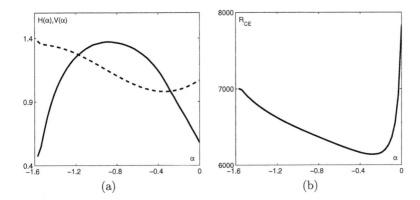

Fig. 2.12 (a) The H_S (solid line), V (dashed line) curves for $a = 0.95$, $b = -1.7$, and $c = 0.9$. (b) The R_{CE} curve for $a = 0.95$, $b = -2.4$, and $c = 0.9$. The curves were obtained by numerical simulation with 4000 instances per class.

Chapter 3
MEE with Continuous Errors

The present chapter analyzes the behavior of classifiers characterized by continuous distributions of the errors, which are trained to minimize error-entropy functionals, namely the Shannon and Rényi's quadratic entropies, presented in the preceding chapter. The analysis focus mainly the classifier problem (does the MEE solution correspond to the min P_e solution?), but consistency and generalization issues are also addressed.

We only consider simple classifiers, used as building blocks of more sophisticated ones. These classifiers are restricted to solving two-class problems; as seen in Sect. 2.2.1, the corresponding univariate error PDF is then expressed as follows for class targets in $\{-1, 1\}$:

$$f_E(e) = p f_{Y|1}(1 - e) + q f_{Y|-1}(-1 - e) . \tag{3.1}$$

Note that the two-class restriction doesn't amount to a loss of generality because all multiclass problems are solved in practice by applying a battery of two-class solvers either in parallel or sequentially.

As we shall soon see, error entropy formulas become rather involved even for simple classifiers applied to simple datasets. Derivation of MEE solutions by some kind of direct parameter estimation method is out of the question. Instead, one has to resort to iterative optimization algorithms. We will use one such algorithm, and a simple one, to analyze the practical behavior of these machines: the gradient descent algorithm. It is simple enough, facilitates the inspection of convergent behavior, and with proper care affords global minima solutions.

J. Marques de Sá et al.: Minimum Error Entropy Classification, SCI 420, pp. 41–91.
springerlink.com

3.1 Theoretical and Empirical MEE

3.1.1 Computational Issues

As with any other risk functional, the practical application of the MEE approach relies on using adequate estimates of the EE risks (2.36), (2.37) and (2.38). This implies using estimates of the error density, based on the n error values produced by the classifier. Given an i.i.d. sample $(x_1, \ldots, x_n)$ from some continuous distribution with density $f(x)$, its estimate can be obtained in an efficient way by the Parzen window method (see Appendix E), which produces the estimate

$$\hat{f}(x) \equiv \hat{f}_n(x) = \frac{1}{n} \sum_{j=1}^{n} \frac{1}{h} K\left(\frac{x - x_j}{h}\right) = \frac{1}{n} \sum_{j=1}^{n} K_h(x - x_j) \,. \tag{3.2}$$

This estimate is also known as *kernel density estimate* (KDE). Properties and optimal choice of the *bandwidth* h for a *kernel* function K are discussed in Appendix E, where the justification to use the Gaussian kernel $G_h(x) = \exp(-x^2/2h^2)/(\sqrt{2\pi}h)$ is also provided.

Applying (3.2) with a Gaussian kernel to the error values e_i, one obtains estimates

$$\hat{f}(e_i) = \frac{1}{n} \sum_{j=1}^{n} G_h(e_i - e_j) \,, \tag{3.3}$$

allowing the estimation of the error entropies as follows:

$$H_S(E) = \mathbb{E}[-\ln f(e)] : \quad \hat{H}_S(E) = -\frac{1}{n} \sum_{i=1}^{n} \ln f(e_i) \approx -\frac{1}{n} \sum_{i=1}^{n} \ln \hat{f}(e_i) \,, \tag{3.4}$$

$$H_{R_2}(E) = -\ln \mathbb{E}[f(e)] : \quad \hat{H}_{R_2}(E) = -\ln \frac{1}{n} \sum_{i=1}^{n} f(e_i) \approx -\ln \frac{1}{n} \sum_{i=1}^{n} \hat{f}(e_i) \,. \tag{3.5}$$

Formulas (3.4) and (3.5) of $\hat{H}_S$ and $\hat{H}_{R_2}$ are plug-in (or resubstitution) estimates of H_S and H_{R_2} (see Appendix F). The minimization of these (or other) empirical EEs corresponds to the *empirical MEE* approach used in practical applications.

From the theoretical point of view, we are interested in analyzing the behavior of the *theoretical MEE*, which for the Shannon and Rényi's quadratic entropies corresponds to the minimization of the risks by formulas (2.36) and (2.38). Unfortunately, theoretical EEs can only be analyzed based on closed-form expressions in simple classifiers with simple input PDFs, such as uniform or Gaussian. For more realistic settings, involving more complex distributions, one has to resort to numerical simulation, which can be carried out as follows: a) generate a large number of samples for each class-conditional input PDF; b) compute the output $\hat{f}_{E|t}(e)$ PDFs with the KDE method; c) compute H_S and H_{R_2} using formulas (C.3) and (C.5) (entropy partitioning) and performing numerical integration. Although laborious, this procedure is able to produce accurate estimates of the theoretical EEs if one uses a "large number of samples" in step (a), that is a value of n in formula (3.2) guaranteeing a very low integrated mean square error (say, $IMSE < 0.01$) of $\hat{f}_{E|t}(e)$ when computed with the optimal $h \equiv h(n)$ (see Appendix E). In these conditions $\hat{f}_{E|t}(e)$ is very close to $f_{E|t}(e)$.

The main differences between empirical and theoretical MEE are as follows:

- Whereas theoretical MEE implies the separate evaluation of $f_{E|t}(e)$, the empirical MEE relies on the estimate of the *whole* $\hat{f}_E(e) \equiv \hat{f}_n(e)$ based on the n-sized dataset.
- One *cannot* apply iterative optimization algorithms to theoretical EEs (at each training step the $f_{E|t}(e)$ are not easily computable); one may, however, compute the theoretical EE in a neighborhood of a parameter vector, as we will do later.
- Whereas the kernel smoothing effect (see Appendix E) is neglectable when using the optimal $h(n)$ in the computation of theoretical EEs, its influence will be of importance, as we shall see, in empirical MEE.

We saw in Sect. 2.2.1 that the $\min P_e = 0$ case for a classifier with interval codomain corresponds to an error PDF with a null area in a subset of the codomain. For instance, for the $[-1, 1]$ codomain and the usual thresholding function assigning the -1 label to negative outputs and the 1 label to positive outputs, the classifier has $P_e = 0$ if the error variable is zero in $[-2, -1] \cup [1, 2]$. In this case the task of the training algorithm is to squeeze $f_E(e)$ (or more rigorously, $\hat{f}_n(e)$ for an arbitrarily large n) driving it inside $[-1, 1]$. Classifiers trained to reach the minimum error entropy should ideally, for infinitely separated classes, achieve more than that: $f_E(e)$ should be driven towards a single Dirac-δ at the origin (see 2.3.1). For other codomain intervals the same result for MEE trained classifiers should ideally be obtained. One question that arises in practical terms is, when we have all n error samples with equal value, $e_1 = \ldots = e_n$, and use the empirical EEs (3.4) and (3.5) instead of the theoretical EEs, are they still a minimum? In other words, do empirical EEs preserve the minimum property of theoretical EEs? The answer is affirmative: the condition $e_1 = \ldots = e_n = 0$ is a minimum, and in fact a global minimum, of the empirical Shannon EE [212].

3.1.2 Empirical EE Gradient Descent

The practical implementation of gradient descent for a classifier with EE risks is described by the following algorithm.

Algorithm 3.1 — EE Gradient descent algorithm

1. For a randomly chosen initial parameter vector, $\mathbf{w}$ (with components w_k), compute the error PDF estimate using the classifier output at the n available data instances, $y_i = \varphi(\mathbf{x}_i; \mathbf{w})$, $i = 1, \ldots, n$.
2. Find the partial derivatives of the empirical EE risks with respect to the parameters. For the empirical Shannon's entropy, one has [212] from formula (3.4):

$$\frac{\partial \hat{H}_S}{\partial w_k} = -\frac{1}{n} \sum_{i=1}^{n} \frac{\partial}{\partial w_k} \ln \hat{f}(e_i) = -\frac{1}{n} \sum_{i=1}^{n} \frac{1}{\hat{f}(e_i)} \frac{\partial \hat{f}(e_i)}{\partial w_k} \ . \tag{3.6}$$

Applying (3.2), assuming a Gaussian kernel (see also (F.2) and (F.3)):

$$\frac{\partial \hat{f}(e_i)}{\partial w_k} = -\frac{1}{n} \sum_{j=1}^{n} \frac{\partial}{\partial w_k} G_h(e_i - e_j)$$

$$= -\frac{1}{nh} \sum_{j=1}^{n} \frac{1}{h^2} G_h(e_i - e_j)(e_i - e_j) \left[\frac{\partial e_i}{\partial w_k} - \frac{\partial e_j}{\partial w_k} \right] . \tag{3.7}$$

Hence,

$$\frac{\partial \hat{H}_S}{\partial w_k} = \frac{1}{n^2 h^3} \sum_{i=1}^{n} \sum_{j=1}^{n} \frac{G_h(e_i - e_j)}{\hat{f}(e_i)} (e_i - e_j) \left[\frac{\partial e_i}{\partial w_k} - \frac{\partial e_j}{\partial w_k} \right] . \tag{3.8}$$

For Rényi's quadratic entropy, one maximizes the empirical information potential and obtains from formula (F.9) [198]:

$$\frac{\partial \hat{V}_{R_2}}{\partial w_k} = -\frac{1}{2n^2 h^2} \sum_{i=1}^{n} \sum_{j=1}^{n} G_{\sqrt{2}h}(e_i - e_j)(e_i - e_j) \left[\frac{\partial e_i}{\partial w_k} - \frac{\partial e_j}{\partial w_k} \right] . \tag{3.9}$$

Note that the only relevant difference between (3.8) and (3.9) is the division by $\hat{f}(e_i)$ in (3.8).

3. Update at each iteration, m, the parameters $w_k^{(m)}$ using a η amount (learning rate) of the gradient. For the Shannon entropy one uses gradient descent:

$$w_k^{(m)} = w_k^{(m-1)} - \eta \frac{\partial \hat{H}_S}{\partial w_k} \Big|_{w_k^{(m-1)}} . \tag{3.10}$$

For the information potential one may use gradient ascent:

$$w_k^{(m)} = w_k^{(m-1)} + \eta \frac{\partial \hat{V}_{R_2}}{\partial w_k} \bigg|_{w_k^{(m-1)}} . \tag{3.11}$$

One may, of course, also apply gradient descent to $-\hat{V}_{R_2}$.

4. With the new parameter vector compute the updated classifier outputs $y_i = \varphi(\mathbf{x}_i; \mathbf{w})$ and go to step 1 if some stopping condition is not met. $\square$

Note that the gradient descent algorithm for EE risks has a $O(n^2)$ complexity, which is a disadvantage relatively to MSE and CE risks with $O(n)$ complexity.

3.1.3 Fat Estimation of Error PDF

We open the present section with two simple examples of PDFs being iteratively driven to the theoretical MEE configuration. In spite of their simplicity these examples serve well as forerunners of things to come.

Example 3.1. Consider the parametric PDF family

$$f(x; d, \sigma) = \frac{1}{2}(g(x; -d, \sigma) + g(x; d, \sigma)) , \tag{3.12}$$

where $g(\cdot)$ is the Gaussian PDF with parameter vector $\mathbf{w} = [d \; \sigma]^T$. We want to study the entropy minimization of the family. We are namely interested, in the present and following companion example, to interpret the PDF family as an error PDF expressed as:

$$f_E(e) = \frac{1}{2}(f_{E|-1}(e) + f_{E|1}(e)) = \frac{1}{2}(g(e; -d, \sigma) + g(e; d, \sigma)) . \tag{3.13}$$

The class-conditional PDFs are equal-variance Gaussian PDFs and the priors are equal ($p = q = 1/2$).

Note that this $f_E(e)$ corresponds to a classifier whose codomain is not a bounded interval as in 2.2.1. Nevertheless, the requirement of driving $f_E(e)$ inside $[-1, 1]$ in order to reach a $\min P_e = 0$ configuration (if possible) still holds. Driving $f_E(e)$ inside $[-1, 1]$ means decreasing both d and σ in such a way that the $f_E(e)$ area outside $[-1, 1]$ is neglectable, and we are therefore arbitrarily close to the $\min P_e = 0$ configuration. Asymptotically, $\mathbf{w}$ would converge to $[0 \; 0]^T$ and $f_E(e)$ towards a single Dirac-δ function at the origin.

We take the Rényi's quadratic entropy as risk functional (greatly facilitating the computations in comparison with Shannon entropy):

$$H_{R_2}(E) = -\ln \int_{-\infty}^{+\infty} f_E^2(e) de = -\ln \int_{-\infty}^{+\infty} \frac{1}{4}(f_{E|-1}(e) + f_{E|1}(e))^2 de ; \tag{3.14}$$

$$V_{R_2}(E) = \exp(-H_{R_2}(E)) = \frac{1}{4} \int_{-\infty}^{+\infty} (f_{E|-1}(e) + f_{E|1}(e))^2 de \ . \qquad (3.15)$$

Applying Theorem F.1 (Appendix F), we have

$$\int_{-\infty}^{+\infty} f_{E|-1}^2(e)de = \int_{-\infty}^{+\infty} f_{E|1}^2(e)de = g(0; 0, \sqrt{2}\sigma) \ ;$$

$$\int_{-\infty}^{+\infty} f_{E|-1}(e)f_{E|1}(e)de = g(0; -2d, \sqrt{2}\sigma) \ . \qquad (3.16)$$

Hence:

$$V_{R_2}(\mathbf{w}) = \frac{1}{4\sqrt{\pi}\sigma}\left(1 + \exp\left(-\frac{d^2}{\sigma^2}\right)\right) \ . \qquad (3.17)$$

As seen in the preceding section, we may apply to V_{R_2} a training algorithm of maximum ascent, using (3.11) with

$$\frac{\partial V_{R_2}}{\partial \sigma} = \frac{1}{4\sqrt{\pi}\sigma^2}\left[\left(\frac{2d^2}{\sigma^2} - 1\right)\exp\left(-\frac{d^2}{\sigma^2}\right) - 1\right] \ ; \qquad (3.18)$$

$$\frac{\partial V_{R_2}}{\partial d} = -\frac{d}{2\sqrt{\pi}\sigma^3}\exp\left(-\frac{d^2}{\sigma^2}\right) \ . \qquad (3.19)$$

Near the maximum of V_{R_2}, in order to prevent an overshoot with a consequent decrease of V_{R_2}, one should decrease η until obtaining an increased V_{R_2}.

Let us select $d = 1$ and $\sigma = 0.9$ as initial parameter values; Fig. 3.1a shows the initial $f_E(e)$. The following Figs. 3.1b through 3.1d show the result, at iterations 28, 30 and 31, of applying gradient ascent to the information potential with $\eta = 0.1$.

There is a clear convergence of $f_E(e)$ towards a single Dirac-δ at the origin. (In fact, towards two infinitely close Dirac-δ functions; the distinction from the single Dirac-δ is unimportant.) Iteration 31 outputs $d = \sigma = 0.19$ with $P_e \approx 0$. The following iteration 32 produces the same Dirac-δ-type solution with smaller magnitudes of d and σ, but demands an overshoot correction with a decrease of η; this behavior repeats itself for following iterations. The converging behavior is essentially the same for other values of the initial learning rate.

We now select as initial parameter values the same $d = 1$ but a smaller $\sigma = 0.4$; the initial $f_E(e)$ is shown in Fig. 3.2a. Gradient ascent with $\eta = 0.01$ [1] produces the evolution shown by Figs. 3.2b through 3.2d. The convergence is now towards two Dirac-δ functions, one very close to -1 the other very close to 1 $(d = 0.9997)$. A small departure from the asymptotic solution means in

---

[1] The value of η influences the convergence rate. The present $\eta = 0.01$ value was chosen so that a convenient number of illustrative intermediary PDFs were obtained.

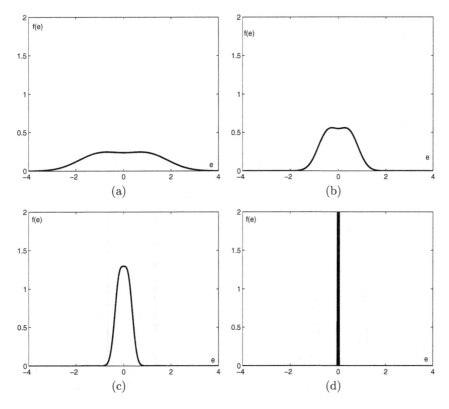

Fig. 3.1 An error PDF (a) converging to a Dirac-δ by gradient ascent of the information potential with $\eta = 0.1$; (b), (c) and (d), show the PDF at iterations 28, 30, and 31, respectively.

this case $P_e \neq 0$. Moreover, note that computer rounding of d to 1 means, in terms of the thresholded classifier outputs (see 2.2.1), that the two-Dirac-δ solution strictly implies $P_e = 0.5$! This effect is felt for small values of the initial σ, $\sigma_{initial}$, (the asymptotic d is 1 with five significant digits for $\sigma_{initial} \lesssim 0.32$). In short, minimization of the error entropy behaves poorly in this case.   □

Example 3.1 illustrates the minimization of theoretical error entropy leading to very different results, depending on the initial choice of parameter vectors. In this example the initial $[1 \; 0.9]^T$ and $[1 \; 0.4]^T$ parameter vectors influenced how the (d, σ) space was explored, in such a way that they led to radically different MEE solutions. The basic explanation for this fact was already given in Sect. 2.3.1: any Dirac-δ comb will achieve minimum entropy (be it Rényi's quadratic or Shannon's), therefore what really matters in Example 3.1 is achieving $\sigma = 0$; the value of d is much less important. One may also see both final solutions as an example of Property 1 mentioned in Sect. 2.3.4.

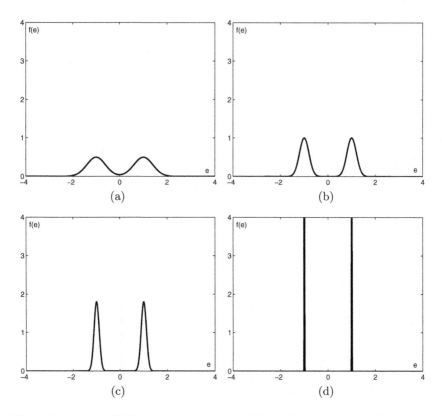

Fig. 3.2 An error PDF a) converging to two Dirac-δ functions by gradient ascent of the information potential with $\eta = 0.01$; b), c) and d), show the error function at iterations 14, 16, and 17, respectively.

(But not before convergence has been achieved, since the supports of the Gaussians are not disjoint.)

We know that the error PDF parameters are directly related to classifier parameters; we may then expect that the choice of initial values in the parameter space of the classifier may drive the learning algorithm towards distinct theoretical MEE solutions. Example 3.1 also suggests that a large $\sigma_{initial}$ tends to drive the MEE solution to the desired min P_e value. We make use of this suggestion in the following example, where we also use the following result:

Proposition 3.1. *The convolution of two Gaussian functions is a Gaussian function whose mean is the sum of the means and whose variance is the sum of the variances:*

$$g(x; \mu_1, \sigma_1) \otimes g(x; \mu_2, \sigma_2) = g(x; \mu_1 + \mu_2, \sqrt{\sigma_1^2 + \sigma_2^2}) \; . \tag{3.20}$$

For a proof, see e.g. [15].

Example 3.2. Let us suppose that the error PDF of Example 3.1 is convolved with $\psi(e) = g(e; 0, h)$ prior to the maximization of the information potential:

$$\frac{1}{2}(f_{E|-1}(e) + f_{E|1}(e)) \otimes \psi(e) \ . \tag{3.21}$$

Since convolution is a linear operator, we can write

$$\frac{1}{2}(f_{E|-1}(e) \otimes \psi(e) + f_{E|1}(e) \otimes \psi(e)) \ . \tag{3.22}$$

We now apply to each $f_{E|t}(e) \otimes \psi(e)$, $t \in \{-1, 1\}$, the result of Proposition 3.1: $f_{E|t}(e) \otimes \psi(e) = g(e; td, \sqrt{\sigma^2 + h^2})$.

Proceeding to the computation of the information potential along the same steps as in the previous example, we finally arrive at

$$V_{R_2} = \frac{1}{4\sqrt{\pi}\sqrt{\sigma^2 + h^2}} \left(1 + \exp\left(-\frac{d^2}{\sigma^2 + h^2}\right)\right) \ . \tag{3.23}$$

Denoting $A = \exp(-d^2/(\sigma^2 + h^2))$, the first-order derivatives

$$\frac{\partial V_{R_2}}{\partial \sigma} = \frac{\sigma}{4\sqrt{\pi}(\sigma^2 + h^2)^{3/2}} \left(-1 - A + \frac{2d^2 A}{\sigma^2 + h^2}\right)$$

and $\tag{3.24}$

$$\frac{\partial V_{R_2}}{\partial d} = -\frac{2dA}{4\sqrt{\pi}(\sigma^2 + h^2)^{3/2}}$$

can be used in (3.11) to perform the gradient ascent maximizing the potential V_{R_2}.

With the derivatives (3.24) one obtains, for sufficiently large h, a convergent behavior similar to the one shown in Fig. 3.1. As a matter of fact, it is enough to use $h \geq 0.75$ to obtain a convergence of both d and σ to zero, and therefore to min $P_e = 0$, even when the initial value of σ is very small (say, 0.01). It is, however, inconvenient to use a too large h, since for the same η the number of needed iterations until reaching some target d (or σ) grows with h according to a cubic law. □

Example 3.2 suggests how to use empirical EEs to our advantage.

For a given n, let $\hat{f}_n(e)$ denote the Parzen window estimate of the error density obtained with the optimal bandwidth in the *IMSE* sense, h_{IMSE}, for that n (see Appendix E):

$$\hat{f}_n(e) = \mu_n \otimes G_{h_{IMSE}}(e) \ , \tag{3.25}$$

where $\mu_n(e)$ is the empirical density of the error. Instead of $\hat{f}_n(e)$, one computes a *fat estimate* $\hat{f}_{fat}(e)$:

$$\hat{f}_{fat}(e) = \mu_n \otimes G_{h_{fat}}(e), \quad \text{with} \quad h_{fat} > h_{IMSE} \ . \tag{3.26}$$

By Proposition 3.1, there is a Gaussian kernel $G_{\tilde{h}}(e)$ such that $G_{h_{fat}}(e) = G_{h_{IMSE}}(e) \otimes G_{\tilde{h}}(e)$. Hence,

$$\hat{f}_{fat}(e) = \mu_n \otimes G_{h_{IMSE}}(e) \otimes G_{\tilde{h}}(e) = \hat{f}_n(e) \otimes G_{\tilde{h}}(e) \xrightarrow[n \to \infty]{} f(e) \otimes G_{\tilde{h}}(e) \ , \tag{3.27}$$

where the convergence is in the IMSE sense.

The estimate $\hat{f}_{fat}(e)$ is oversmoothed compared to the one converging to $f(e)$. This is unimportant since we are not really interested in $\hat{f}_n(e)$ (we namely don't use it to compute error rates). Our sole interest is in getting the right classifier parameter values (d and σ in Examples 3.1 and 3.2) corresponding to min P_e.

3.2 The Linear Discriminant

Linear discriminants are basic building blocks in data classification. The linear discriminant implements the following classifier family:

$$\mathcal{Z}_W = \left\{ \theta(\mathbf{w}^T \mathbf{x} + w_0); \mathbf{w} \in W \subset \mathbb{R}^d, w_0 \in \mathbb{R} \right\} \ , \tag{3.28}$$

where $\mathbf{w}$ and w_0 are the classifier parameters usually known as *weight vector* and *bias* term, respectively, and $\theta(\cdot)$ is the usual classifier thresholding function yielding class codes. We restrict here our analysis of the linear discriminant to the case where the inputs are Gaussian distributed; this will be enough to demonstrate the MEE sub-optimal behavior for this type of classifier.

3.2.1 Gaussian Inputs

To derive the error PDF for Gaussian inputs x_i we take into account that Gaussianity is preserved under linear transformations: if X with realizations $\mathbf{x} = [x_1 \ldots x_d]^T$ has a multivariate Gaussian distribution with mean μ and covariance Σ, $X \sim g(\mathbf{x}; \mu, \Sigma)$, then

$$Y = \mathbf{w}^T X + w_0 \sim g(y; \mathbf{w}^T \mu + w_0, \mathbf{w}^T \Sigma \mathbf{w}) \ . \tag{3.29}$$

Therefore, the class-conditional error PDFs, $f_{E|t}(e)$, are also Gaussian and we deal with an error PDF setting similar to the one of Examples 3.1 and 3.2:

$$f_{Y|t}(y) = g(y; \mathbf{w}^T \mu_{X|t} + w_0, \mathbf{w}^T \Sigma_{X|t} \mathbf{w}) \tag{3.30}$$

and

$$f_{E|t}(e) = f_{Y|t}(t - e) = g(e; t - \mathbf{w}^T \boldsymbol{\mu}_{X|t} - w_0, \mathbf{w}^T \boldsymbol{\Sigma}_{X|t} \mathbf{w}) . \tag{3.31}$$

We now proceed to compute the information potential as in Example 3.1:

$$\int_{-\infty}^{+\infty} f_{E|t}^2(e) de = g(0; 0, \sqrt{2}\sigma_{Y|t}) \quad \text{and}$$

$$\int_{-\infty}^{+\infty} f_{E|-1}(e) f_{E|1}(e) de = g(0; 2 - d, \sqrt{2}\sigma_m) , \tag{3.32}$$

with $\sigma_{Y|t} = \sqrt{\mathbf{w}^T \boldsymbol{\Sigma}_{X|t} \mathbf{w}}$, $\sigma_m = \sqrt{\sigma_{Y|-1}^2 + \sigma_{Y|1}^2}$, and $d = \mathbf{w}^T(\boldsymbol{\mu}_{X|1} - \boldsymbol{\mu}_{X|-1})$.
Hence, for equal priors:

$$V_{R_2} \equiv V_{R_2}(d, \sigma_{Y|-1}, \sigma_{Y|1}) =$$

$$\frac{1}{8\sqrt{\pi}} \left(\frac{1}{\sigma_{Y|-1}} + \frac{1}{\sigma_{Y|-1}} + \frac{2}{\sigma_m} \exp\left(-\frac{(2-d)^2}{4\sigma_m^2} \right) \right) . \tag{3.33}$$

It is clear that Rényi's quadratic entropy doesn't depend on w_0. This is a direct consequence of the invariance of entropy to translations, since from (3.31) we observe that $\mathbb{E}[E] = -\mathbf{w}^T \boldsymbol{\mu}_{X|-1} - \mathbf{w}^T \boldsymbol{\mu}_{X|1} - w_0$. Shannon's entropy and α-order Rényi's entropies are insensitive to the constant w_0 term.

Things are different, however, when a linear classifier is trained with gradient descent using empirical entropies. Off the convergent solution, the $e_i - e_j$ deviations in formula (3.3) are scattered, and the estimate $\hat{f}(e)$ doesn't usually reproduce well a sum of Gaussians with the above mean value. As a consequence, the bias term of the solution will undergo adjustments. Near the convergent solution, with the $e_i - e_j$ deviations crowding a small interval, the $\hat{f}(e)$ estimate then provides a close approximation of the theoretical error PDF and the insensitivity to bias adjustments plays its role.

This empirical MEE behavior is illustrated in the following bivariate two-class example, where Shannon's entropy gradient descent is used.

Example 3.3. Consider two normally distributed class-conditional PDFs, $g(\mathbf{x}; \boldsymbol{\mu}_t, \boldsymbol{\Sigma}_t)$, with

$$\boldsymbol{\mu}_{-1} = [0 \ 0]^T, \ \boldsymbol{\mu}_1 = [2 \ 0]^T, \ \boldsymbol{\Sigma}_{-1} = \boldsymbol{\Sigma}_1 = \mathbf{I}.$$

Independent training and test datasets with $n = 250$-instances (125 instances per class) were generated and the Shannon MEE algorithm applied with $h = 1$ and $\eta = 0.001$ [2]. Note that according to formula (E.19) the optimal bandwidth for the number of instances being used is $h_{IMSE} = 0.4$. We are, therefore, using fat estimation of the error PDF.

---

[2] From now on the indicated η values are initial values of an adaptive rule to be described in Sect. 6.1.1.

Figure 3.3 shows the dataset (the -1 class corresponds to the right cluster), the evolution of the decision border, and the error PDF (computed only at the e_i values) at successive iteration steps, known as epochs, in one experiment. Epochs $= 0$ is the initial configuration with random weights and bias. Note the evolution of $\hat{f}(e)$ towards a Gaussian-resembling PDF.

The solution at Epochs $= 50$ is practically the same as it was at Epochs $= 35$ and doesn't change thereafter; it is a convergent solution. The entropy and error rate graphs also become flat after Epochs $= 35$. We see that this convergent solution is somewhat deviated from the optimal solution (the vertical line at $x_1 = 1$) because of insufficient bias adjustment in the last iterations.

The min P_e value is exactly known for the two-class setting with equal-covariance Gaussian inputs, separated by a linear discriminant. The min P_e is then also the Bayes optimal error given by [76]

$$P_{e_{Bayes}} = 1 - \Phi(\delta/2) , \tag{3.34}$$

with $\Phi(\cdot)$ the standardized normal CDF and $\delta^2 = (\mu_{-1} - \mu_1)^T \Sigma^{-1} (\mu_{-1} - \mu_1)$, the Mahalanobis distance. In the present case $P_{e_{Bayes}} = 0.1587$.

We see that the final solution exhibits a training set and test set error of around 0.23. However, at epochs $= 28$ a solution closer to the optimal one, with training set error $\hat{P}_{ed} = 0.166$ and test set error $\hat{P}_{et} = 0.164$ had been reached; progress from this stage was hindered by the insensitivity to w_0. □

Example 3.4. This example is similar to the preceding one; the only difference is that we now use $h = 0.4$, as predicted by the optimal IMSE formula (E.19). The experiment illustrated in Fig. 3.4 clearly shows the convergence to two Dirac-δ functions near 1 and -1, as in Example 3.1 when an initial small σ was used. This should not be surprising since σ is here $\sqrt{\mathbf{w}^T \mathbf{w}}$, implying a convergence of the weights to zero, therefore to an error PDF represented by two Dirac-δ functions at $\pm 1 + w_0$, with the bias w_0 representing here the same role as parameter d in Example 3.1. As a consequence we obtain a behavior already described in Sect. 3.1.3 with a final $\hat{P}_{ed} = \hat{P}_{et} = 0.5$. Note that with zero weights the decision border is undefined. □

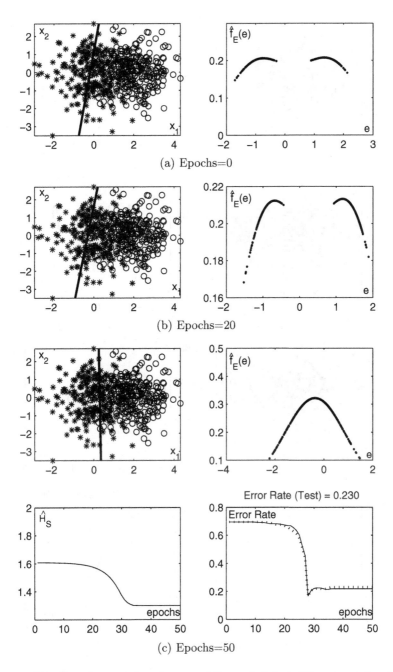

(a) Epochs=0

(b) Epochs=20

(c) Epochs=50

Error Rate (Test) = 0.230

Fig. 3.3 The Gaussian two-class dataset of Example 3.3 at different epochs of the MEE linear discriminant. The upper three graphs at the left side show the dataset with the linear decision border (solid line); the graphs at the right side show the error PDF. The bottom graphs show the evolution of the Shannon entropy and the error rate for the training set (solid) and test set (dotted).

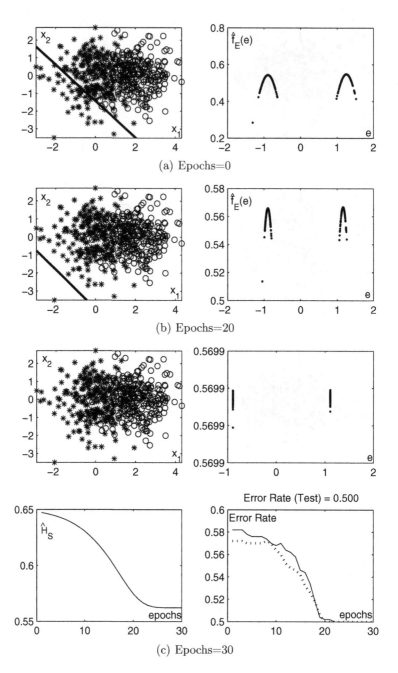

(a) Epochs=0

(b) Epochs=20

(c) Epochs=30

Fig. 3.4 Graphs showing an experiment as in Fig. 3.3 but with the error PDFs estimated with $h = 0.4$ (no fat estimation).

Example 3.5. Again an example similar to Example 3.3 (fat estimation of the error PDF); the only difference is that we now use smaller variances: $\Sigma_{-1} = \Sigma_1 = 0.12 \times \mathbf{I}$, drastically increasing the separability of the classes. In this case $\min P_e = P_{e_{Bayes}} \approx 0$.

Figure 3.5 again shows a convergent solution that is sub-optimal due to lack of bias adjustment. Interestingly enough at epochs 23 and 24 the test set error rate was zero. The linear discriminant moved from right to left and completely separated the classes (zero training set error) at epoch 24, when the Shannon entropy was still decreasing. The bias adjustment was then lost and the discriminant still moved slightly to the left pulled by other parameter adjustments until the entropy stabilized near epoch 32.   □

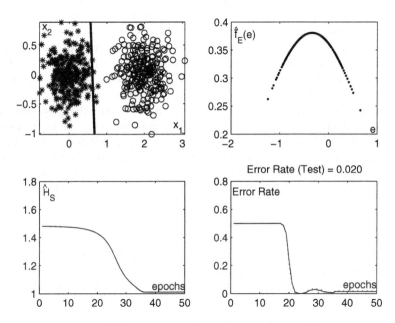

Fig. 3.5 Graphs showing an experiment as in Fig. 3.3 (epochs=50), but with well-separated classes.

3.2.2 Consistency and Generalization

The one-experiment runs in the preceding section help to get insight into MEE trained linear discriminants, but are of course insufficient to draw general conclusions, even when we stick to the two-class bivariate Gaussian input scenario.

Let us consider a classifier designed by empirical risk minimization using some (X_{ds}, T_{ds}) training set. We now denote by n_d the number of (X_{ds}, T_{ds}) instances. The classifier error rate estimated by error counting in (X_{ds}, T_{ds}) (empirical error formula (1.4)) is then denoted $\hat{P}_{ed}(n_d)$. The generalization ability of the classifier is assessed by measuring its error rate in some n_t-sized test set, (X_{ts}, T_{ts}), assumed to be independently drawn from the same $F_{X,T}$ distribution; the respective error rate is $\hat{P}_{et}(n_t)$. $\hat{P}_{ed}(n_d)$ and $\hat{P}_{et}(n_t)$ are single realizations of two r.v.s; we are then interested in the main statistics of these r.v.s, namely their expected values characterizing the average training set and test set behaviors of the classifier. Usually, there are no closed-form expressions of these expected values (an exception is the two-class setting with equal-variance Gaussian inputs; see [74]) and one has to use sample mean estimates, which we denote $\overline{P}_{ed}(n_d)$ and $\overline{P}_{et}(n_t)$, respectively.

If, for some classifier designed with n_d instances, one computes $\overline{P}_{et}(n_t)$ with a sufficiently large number of (X_{ts}, T_{ts}) sets s.t. $\overline{P}_{et}(n_t) \approx \mathbb{E}[\hat{P}_{et}(n_t)]$, then $\overline{P}_{et}(n_t)$ is also a close estimation of $P_e(n_d)$ the true error of the classifier (see e.g., [76]). On the other hand, for a sufficiently large number of (X_{ds}, T_{ds}) sets, $\overline{P}_{ed}(n_d) \approx \mathbb{E}[\hat{P}_{ed}(n_d)]$ will exhibit a bias monotonically decreasing with n_d relative to $\min P_e$, for consistent learning algorithms.

Unfortunately, in the usual practice one doesn't have the possibility of drawing large numbers of independent training sets and test sets. In the usual practice, one is given an n-sized (X_{dt}, T_{dt}) set which will be used for both classifier design and classifier evaluation. One is then compelled to partition (X_{dt}, T_{dt}) into training sets and test sets. There are a number of ways of randomly selecting n_d and n_t instances for the computation of $\overline{P}_{ed}(n_d)$ and $\overline{P}_{et}(n_t)$. The basic scheme we use throughout the book is the *stratified k-fold cross-validation* scheme, where the n-sized (X_{dt}, T_{dt}) dataset is first randomly partitioned into k partitions of size $n_t = n/k$, such that the classes maintain their proportional representation (stratification). One is then able to carry out k design experiments, using each time all the instances of $(k-1)$ different partitions as design set and testing the classifier in the remaining partition. Training set and test set averages $(\overline{P}_{ed}(n_d), \overline{P}_{et}(n_t))$ and standard deviations $(s(P_{ed}(n_d)), s(P_{et}(n_t)))$ of the k experiments are then computed. The standard deviations are always decreasing with the set cardinalities (as in the binomial approximation $\sqrt{P_e(1 - P_e)/n}$). Note that the partition scheme invalidates the assumption of test sets being independent of training sets. The practical implications of this, with respect to what we said above about $\overline{P}_{et}(n_t)$ and $\overline{P}_{ed}(n_d)$, are not important for not too small n.

Sometimes, the k-fold cross-validation scheme is repeated a certain number of times, each time with a (X_{ds}, T_{ds}) reshuffling, with improvement of the quality (confidence interval) of the estimates.

The particular case of the cross-validation scheme with $k = 2$ is called *hold-out* method. We use in the following examples a simpler version of the repeated hold-out method (*simple hold-out*) where half of the data is used for training and the other half for test (without swapping their roles). We assume the availability of $2n$-sized (X_{dt}, T_{dt}) sets allowing the computation of $\overline{P}_{ed}(n)$, $\overline{P}_{et}(n)$ by simple hold-out for increasing n. A consistent learning algorithm of classifier design will exhibit $\overline{P}_{ed}(n)$, $\overline{P}_{et}(n)$ curves — *learning curves* — converging to the $\min P_e$ value.

Example 3.6. We consider the same classifier problem as in Example 3.3. The results of that example lead us to suspect a convergence of P_e towards a value above 0.2.

We now study in more detail the convergence properties of the MEE linear discriminant for this dataset, by performing 25 experiments for n from 5 to 195 with increments of 10, using simple holdout. The $\hat{P}_{ed}(n)$ and $\hat{P}_{et}(n)$ statistics are then computed for the 25 experiments.

Figure 3.6 shows the learning curves $\overline{P}_{ed}(n) \pm s(P_{ed}(n))$ and $\overline{P}_{et}(n) \pm s(P_{et}(n))$, illustrating two important facts:

1. There is a clear convergence of both $\overline{P}_{ed}(n)$ and $\overline{P}_{et}(n)$ towards the same asymptotic value $P_e = 0.237$. However, learning is *not* consistent in the $\min P_e$ (0.1587) sense. As usual, the convergence of $\overline{P}_{ed}(n)$ is from below (the training set error rate is optimistic on average) and the convergence of $\overline{P}_{et}(n)$ is from above (the test set error rate is pessimistic on average).
2. From very small values of n (around 50) onwards the $\overline{P}_{et}(n) - \overline{P}_{ed}(n)$ difference is small. The MEE linear discriminant *generalizes well* for this dataset.

$\square$

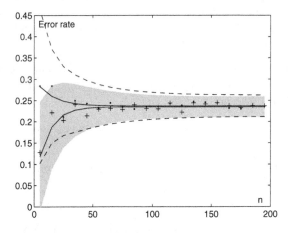

Fig. 3.6 Learning curves for the MEE linear discriminant applied to the Example 3.3 dataset. The learning curves (solid lines) were obtained by exponential fits to the $\overline{P}_{ed}$ (denoted '+') and $\overline{P}_{et}$ (denoted '.') values. The shadowed region represents $\overline{P}_{ed} \pm s(P_{ed})$; the dashed lines represent $\overline{P}_{et} \pm s(P_{et})$.

The asymptotic and generalization behaviors of Example 3.6 can be confirmed for other Gaussian datasets with equal and unequal covariances. There is a theoretical justification for the good generalization of the MEE linear discriminant with independent Gaussian inputs. It is based on the following

Theorem 3.1. *The minimization of Shannon's or Rényi's quadratic entropy of a weighted sum of d independent Gaussian distributions implies the minimization of the norm of the weights.*

Proof. The weighted sum of d independent Gaussian distributions $y = \mathbf{w}^T \mathbf{x}$ has the PDF

$$f(y) = g(y; \mathbf{w}^T \boldsymbol{\mu}, \mathbf{w}^T \boldsymbol{\Sigma} \mathbf{w}) , \qquad (3.35)$$

with $\boldsymbol{\Sigma}$ a diagonal matrix of the variances since the distributions are independent. But:

$$H_S(Y) = \ln\left(\sqrt{2\pi e \mathbf{w}^T \boldsymbol{\Sigma} \mathbf{w}} \right) ; \qquad (3.36)$$

$$H_{R_2}(Y) = \ln\left(2\sqrt{\pi \mathbf{w}^T \boldsymbol{\Sigma} \mathbf{w}} \right) . \qquad (3.37)$$

The quadratic form $\mathbf{w}^T \boldsymbol{\Sigma} \mathbf{w}$ can be written as $\sum_{i=1}^d w_i^2 \sigma_i^2$; therefore, the minimization of either H_S or H_{R_2} implies the minimization of $\|\mathbf{w}\|^2$. □

Whenever the error PDF approaches a Gaussian distribution in the final stages of the training process, Theorem 3.1 applies and we expect the minimization of $\|\mathbf{w}\|^2$ to take place. As is known from the theory of SVMs, the minimization of $\|\mathbf{w}\|^2$ is desirable since it implies a smaller Vapnik-Chervonenkis distance, therefore smaller classifier complexity with better generalization [228, 43]. As a matter of fact, for Rényi's quadratic entropy a stronger assertion can be made:

Corollary 3.1. *The minimization of Rényi's quadratic entropy of the error of a linear discriminant for independent Gaussian input distributions implies the minimization of the norm of the weights.*

Proof. We have: $f_{Y|t}(y) = g(y; m_t, \sigma_t)$ with $m_t = \mathbf{w}^T \boldsymbol{\mu}_t + w_0$, $\sigma^2 = \mathbf{w}^T \boldsymbol{\Sigma} \mathbf{w}$;

$$f_{E|t}(e) = f_{Y|t}(t - e) = \frac{1}{\sqrt{2\pi}\sigma} \exp\left(-\frac{1}{2\sigma^2}(t - e - m_t)^2 \right) . \qquad (3.38)$$

Therefore:

$$V_{R_2}(E) = \frac{1}{4\sqrt{\pi}\sigma} \left(1 + \exp\left(-\frac{(2 + \mathbf{w}^T(\boldsymbol{\mu}_{-1} - \boldsymbol{\mu}_1))^2}{4\sigma^2} \right) \right) , \qquad (3.39)$$

which is an increasing function for a decreasing σ, thus, as we saw in Theorem 3.1, with decreasing $\|\mathbf{w}\|^2$. □

3.3 The Continuous-Output Perceptron

The continuous-output perceptron (also called *neuron*) is a simple classifier represented diagrammatically in Fig. 3.7.

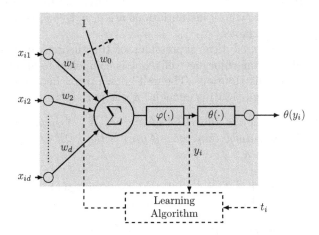

Fig. 3.7 Perceptron diagram (shaded gray) with learning algorithm.

The device implements the following classifier function family:

$$\mathcal{Z}_W = \left\{ \theta(\varphi(\mathbf{w}^T \mathbf{x} + w_0)); \mathbf{w} \in W \subset \mathbb{R}^d, w_0 \in \mathbb{R} \right\} , \qquad (3.40)$$

where $\mathbf{w}$ and w_0 are the classifier parameters known in perceptron terminology as *weights* and *bias*, $\varphi(.)$ is a continuous *activation function*, and $\theta(\cdot)$ the usual classifier thresholding function yielding class codes. In the original proposal by Rosenblatt [190] the perceptron didn't have a continuous activation function.

Note that $\mathcal{Z}_W$ is sufficiently rich to comprehend the function families of linear discriminants and data splitters.

The activation function $\varphi(\cdot)$ is usually some type of squashing function, i.e., a continuous differentiable function such that $\varphi(x) \in [a, b]$, $\lim_{x \to -\infty} \varphi(x) = a$ and $\lim_{x \to +\infty} \varphi(x) = b$. We are namely interested in strict monotonically increasing squashing functions, popularly known as *sigmoidal* (S-shaped) functions, such as the hyperbolic tangent, $y = \tanh(x) = (e^x - e^{-x})/(e^x + e^{-x})$ or the logistic sigmoid $y = 1/(1 + e^{-x})$. These are the ones that have been almost exclusively used, but there is no reason not to use other functions, like for instance the trigonometric arc-tangent function $y = \text{atan}(x)$.

The perceptron implements a linear decision border in d-dimensional space defined by $\varphi(\mathbf{w}^T \mathbf{x} + w_0) = a$ ($a = 0$ for the $\tanh(\cdot)$ sigmoid with $T = \{-1, 1\}$, and $a = 0.5$ for the logistic sigmoid with $T = \{0, 1\}$). For instance, for

$d = 2$ we get a decision surface in three-dimensional space, which for a linear activation function is a plane (the linear discriminant of 3.28) and for a squashing function an S-folded plane; the decision border is always a line.

In the present section we analyze the perceptron as a regression-like machine: the learning algorithm iteratively drives the weights such that the *continuous* output $y_i = \varphi(\mathbf{w}^T \mathbf{x}_i + w_0)$ approximates the target value t_i for every $\mathbf{x}_i$. The error r.v. whose instantiations are $e_i = t_i - y_i$ is, therefore, a continuous random variable.

In order to derive analytical expressions of the theoretical EEs for the perceptron (and other machines as well), one is compelled to apply transformations of the input distributions. The well-known theorem of univariate r.v. transformation (see e.g., [183]) is enough for our purposes:

Theorem 3.2. *Let $f(x)$ be the PDF of the r.v. X. Assume $\varphi(x)$ to be a monotonic and differentiable function. If $g(y)$ is the PDF of $Y = \varphi(X)$ and $\varphi'(x) \neq 0$, $\forall x \in X$, then*

$$g(y) = \begin{cases} \frac{f(\varphi^{-1}(y))}{|\varphi'(\varphi^{-1}(y))|} & \inf \varphi(x) < y < \sup \varphi(x) \\ 0 & otherwise \end{cases} , \qquad (3.41)$$

where $x = \varphi^{-1}(y)$ is the inverse function of $y = \varphi(x)$. $\qquad\square$

Note that sigmoidal activation functions satisfy the conditions of the theorem.

3.3.1 Motivational Examples

Two examples are now presented which constitute a good illustration on how the Shannon EE-risk perceptron, with $\varphi(\cdot) = \tanh(\cdot)$ and $T = \{-1, 1\}$, performs when applied to two-class, two-dimensional datasets. They involve artificial data for which the $\min P_e$ values are known. We will leave the application to real-world datasets to a later section. We will also use these examples to discuss some convergence issues.

Example 3.7. In this example we use two Gaussian distributed class-conditional PDFs, $g(\mathbf{x}; \mu_t, \Sigma_t)$, with

$$\mu_{-1} = [0 \ \ 0]^T, \ \Sigma_{-1} = I, \ \text{and} \ \mu_1 = [1.5 \ \ 0.5]^T, \ \Sigma_1 = \begin{bmatrix} 1.1 & 0.3 \\ 0.3 & 1.5 \end{bmatrix}.$$

Let us consider 300-instance training and test datasets (150 instances per class) and apply the (Shannon) MEE gradient descent algorithm with $h = 1$ and $\eta = 0.001$. Note that according to formula (E.19) the optimal bandwidth for $n = 150$ is $h_{IMSE} = 0.39$. We are, therefore, using fat estimation of the error PDF.

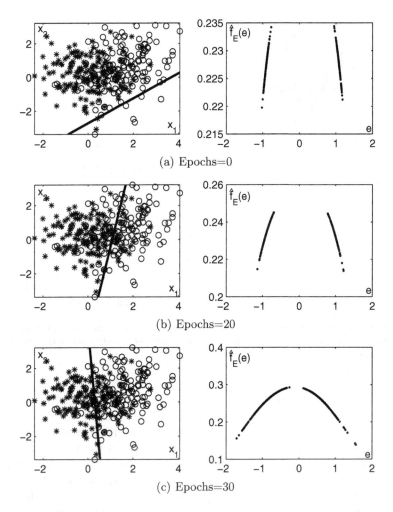

Fig. 3.8 The Gaussian two-class dataset at different iterations of the MEE perceptron. The left graphs show the dataset with the linear decision border (solid line).
The right graphs show the error PDF in the $E = [-2, 2]$ support.

Figure 3.8 shows the evolution of the decision border and the error PDF
(computed only at the e_i values) at successive epochs. Epochs = 0 is the initial
configuration, with random weights and bias. Note the evolution towards a
monomodal error PDF.

Figure 3.9 shows the final solution, corresponding to a converged behavior
of the algorithm to low entropy and error rates. This solution has $P_{ed} = 0.213$
and $P_{et} = 0.227$. The value of the min P_e for this problem is ≈ 0.228. The
$P_{e_{Bayes}}$ is smaller since it is well-known that it can only be achieved with a
quadratic decision border.

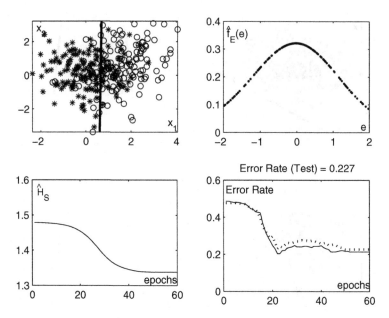

Fig. 3.9 The final converged solution of Example 3.7. The downside graphs of the Shannon entropy and the error rate (solid line for the training set and dotted line for the test set) are in terms of the no. of epochs.

A word about the operational parameter h: had we used a too small h (say, 0.3) we wouldn't have obtained the convergence to the desired solution; instead, we would have got the convergence to two Dirac-δ functions as in Example 3.4. The smallest value of h guaranteeing in the present case the desired convergence is ≈ 0.75. For too high values of h the error PDF is oversmoothed, poorly reflecting the class-conditional components of the error. As a result the error rates behave in a somewhat erratic way as shown in Fig. 3.10 for $h = 3$, with a tendency of poor generalization. In the present case it was found that h should not be higher than 2.5.   □

Example 3.8. The class-conditional distributions of the input data for this example are circular uniform, defined as:

$$f_{X|t}(\mathbf{x}) = cu(\mathbf{x}; \mu_t, r_t) = \begin{cases} \frac{1}{\pi r_t^2} & \|\mathbf{x} - \mu_t)\|^2 \le r_t^2 \\ 0 & \text{otherwise} \end{cases}. \tag{3.42}$$

Let us consider 300-instance training and test datasets (150 instances per class), distributed as in (3.42) with $\mu_{-1} = [0 \ \ 0]^T$, $r_{-1} = 1$ and $\mu_1 = [3 \ \ 0]^T$, $r_1 = 2$. The (Shannon) MEE algorithm was applied with $h = 0.7$ (fat estimation) and $\eta = 0.001$.

Figure 3.11 shows the evolution of the decision border and the error PDF (computed only at the e_i values) in one experiment. Note the two peaks of

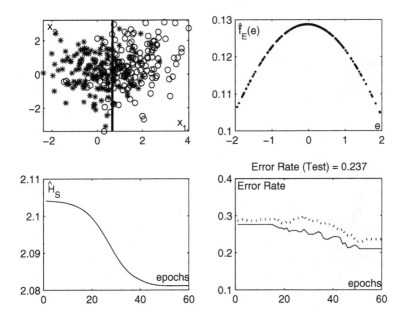

Fig. 3.10 An experiment ran for Example 3.7 with a high value of h: $h = 3$.

the error PDF which merge at a later stage into each other resulting in a monomodal PDF. Figure 3.12 shows the final solution, corresponding to a converged behavior of the algorithm with low entropy and error rates. This solution has $\hat{P}_{ed} = 0$, which is actually the value of the min P_e for this problem and coincident with $P_{e_{Bayes}}$. The test set error rate is also very close to zero.

□

The results of Examples 3.7 and 3.8 exhibit a convergence to a minimum error rate together with a minimum error entropy, but they are just one run of the algorithm; they don't tell us whether the MEE perceptron is learning in a consistent way, and is converging or not towards min P_e. Experimental evidence on these issues can be gained through the learning curves, as we did for the linear discriminant.

Figure 3.13a corresponds to generating 30 times the Gaussian dataset of Example 3.7, for a grid of n values in $[10, 250]$, and computing averages and standard deviations of training set and test set errors ($P_{ed}(n)$ and $P_{et}(n)$) for the 30 repetitions. The maximum number of epochs was 60 and $\eta = 0.001$. Perceptron training in these experiments used Rényi's quadratic EE. Figure 3.13b corresponds to analogous experiments for the circular uniform dataset with parameters as in Example 3.8 except for r_1 which was set to 3. Shannon EE was used in these experiments.

The learning curves $\overline{P}_{ed}, \overline{P}_{et}$ clearly converge with n with decreasing standard deviations as expected. The asymptotic error rates in Fig. 3.13 are clearly very close to the theoretical min P_e values, respectively 0.228 and 0.054.

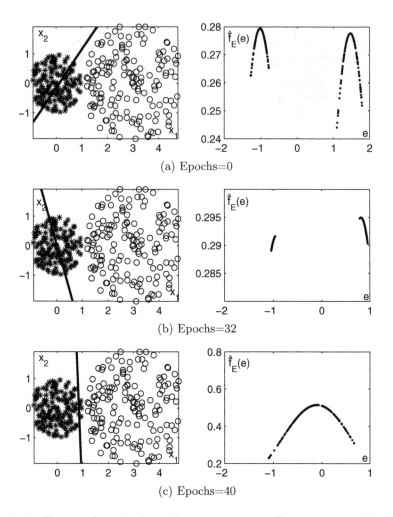

Fig. 3.11 The two-class circular uniform datasets at different epochs of the MEE perceptron. The left graphs show the datasets with the linear decision border (solid line). The right graphs show the error PDF in the $E = [-2, 2]$ support.

Table 3.1 shows other comparisons between the $\min P_e$ and $\overline{P}_{ed}(300)$ values for the two problem types. The $\min P_e$ value for the circular uniform distribution can be obtained by straightforward trigonometric derivations; amazingly, however, the optimal vertical line separating the classes does not correspond, in general, to the intersection point of the circles, and there is no closed-form algebraic expression of its distance from the origin. Its value was determined by optimization search techniques. For the Gaussian distributions there is also, in general, no closed-form expression, except for equal covariance

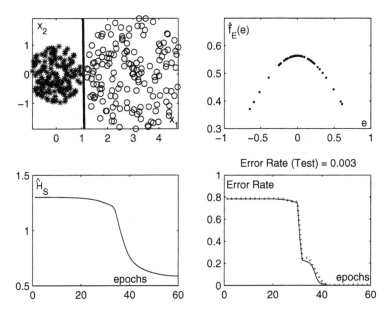

Fig. 3.12 The final converged solution of Example 3.8. The downside graphs of the Shannon entropy and the error rate (solid line for the training set and dashed line for the test set) are in terms of the no. of epochs.

Table 3.1 $\overline{P}_{ed}(300) \pm s(P_{ed}(300))$ and $\min P_e$ values for the distributions of Examples 3.7 and 3.8. Equal priors are assumed.

Circular uniform distributions		
Parameters[a]	$\overline{P}_{ed}(300) \pm s(P_{ed}(300))$	$\min P_e$
$\mu_{11} = 2,\ r_1 = 1.2$	0.0161 ± 0.0071	0.0160
$\mu_{11} = 3,\ r_1 = 2.5$	0.0258 ± 0.0061	0.0253
$\mu_{11} = 3,\ r_1 = 3.0$	0.0549 ± 0.0113	0.0540
Gaussian distributions		
Parameters[b]	$\overline{P}_{ed}(300) \pm s(P_{ed}(300))$	$\min P_e$
$\mu_1 = [2\ 0]^T,\ \Sigma_1 = \mathbf{I}$	0.1595 ± 0.0169	$0.1587^{[c]}$
$\mu_1 = [2\ 0]^T,\ \Sigma_1 = \begin{bmatrix} 1.2 & 0 \\ 0 & 2 \end{bmatrix}$	0.1634 ± 0.0159	≈ 0.1697
$\mu_1 = [1.5\ 0.5]^T,\ \Sigma_1 = \begin{bmatrix} 1.1 & 0.3 \\ 0.3 & 1.5 \end{bmatrix}$	0.2261 ± 0.0156	≈ 0.2281

[a]$\mu_{-1} = [0\ 0]^T$, $r_{-1} = 1$, $\mu_{12} = 0$; [b]$\mu_{-1} = [0\ 0]^T$, $\Sigma_{-1} = \mathbf{I}$; [c] exact value.

matrices as we mentioned in 3.2.1 (formula (3.34)); the work [191] provides, however, for the general case, the means of computing an approximation of $\min P_e$, with a deviation reported to be at most 0.0184.

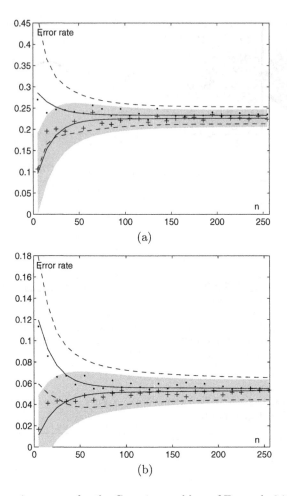

Fig. 3.13 Learning curves for the Gaussian problem of Example 3.7 (bottom) and the circular uniform problem of Example 3.8 with $r_1 = 3$ (top), using respectively the H_{R_2}-MEE and the H_S-MEE perceptron. The learning curves (solid lines) were obtained by exponential fits to the $\overline{P}_{ed}(n)$ (denoted '+') and $\overline{P}_{et}(n)$ (denoted '.') values. The shadowed region represents $\overline{P}_{ed} \pm s(P_{ed})$; the dashed lines represent $\overline{P}_{et} \pm s(P_{et})$.

Similar results were obtained when using different numbers of instances per class, reflecting different values of the priors p and q.

Briefly, these (and other) experiments provide experimental evidence that the (empirical) MEE perceptron learns consistently and converges towards the $\min P_e$ classifier, for the Gaussian and circular uniform input distributions.

3.3.2   Theoretical and Empirical MEE Behaviors

We start by analyzing simple settings with Gaussian inputs and then move on to more realistic settings. The simple Gaussian-input settings provide the basic insights on the distinct aspects of theoretical and empirical MEE, in a theoretically controlled way. More realistic datasets serve to confirm those insights.

3.3.2.1   Univariate and Bivariate Gaussian Datasets

Let us consider the perceptron with Gaussian inputs and the tanh activation function. Applying Theorem 3.2, the class-conditional error densities are:

$$f_{E|t}(e) = \frac{\exp\left(-\frac{1}{2}\frac{(\operatorname{atanh}(t-e)-(\mathbf{w}^T\boldsymbol{\mu}_t+w_0))^2}{\mathbf{w}^T\boldsymbol{\Sigma}_t\mathbf{w}}\right)}{\sqrt{2\pi\mathbf{w}^T\boldsymbol{\Sigma}_t\mathbf{w}}\,e(2t-e)}\mathbb{1}_{]t-1,t+1[}(e)\,. \tag{3.43}$$

We first consider the univariate case $\varphi(w_1x+w_0)$ with w_1 controlling the steepness of the activation function; the error density is then

$$f_{E|t}(e) = \frac{\exp\left(-\frac{1}{2}\frac{(\operatorname{atanh}(t-e)-(w_1\mu_t+w_0))^2}{w_1\sigma_t}\right)}{\sqrt{2\pi}w_1\sigma_t e(2t-e)}\mathbb{1}_{]t-1,t+1[}(e)\,. \tag{3.44}$$

Even for this simple case there is no closed-form expression of H_S (or H_{R_2}). One has to resort to numerical integration and apply expressions (C.3) (or (C.5)). Setting w.l.o.g. $(\mu_{-1},\,\sigma_{-1})=(0,1)$ we obtain the H_S behavior shown in Fig. 3.14 [212].

Figure 3.14a corresponds to $(\mu_1,\,\sigma_1)=(3,1)$. The optimal split point (the "decision border" in this case) is at $x^*=1.5$. We observe that for small values of w_1 (top figure) H_S exhibits a maximum at the optimal split point, instead of a minimum. A minimum is obtained for a sufficiently large w_1 (bottom figure). The same behavior is observed in Fig. 3.14b corresponding to $(\mu_1,\,\sigma_1)=(1,1)$ with $x^*=0.5$. This behavior is, in fact, general for both H_S and H_{R_2}, and no matter the degree of distribution overlap: the theoretical MEE perceptron is able to produce the min P_e solution.

We now move to the bivariate case, fixing $\boldsymbol{\mu}_{-1}=[0\ 0]^T$, $\boldsymbol{\Sigma}_t=\mathbf{I}$, and study two different settings: $\boldsymbol{\mu}_1=[5\ 0]^T$ (distant classes) and $\boldsymbol{\mu}_1=[1\ 0]^T$ (close classes). For these two settings the min P_e value is 0.0062 and 0.3085, respectively. These min P_e values correspond to infinitely many optimal solutions $\mathbf{w}^*=[w_1^*\ 0\ w_0^*]^T$: any (w_1^*,w_0^*) pair s.t. $-w_0^*/w_1^*=2.5$ and $-w_0^*/w_1^*=0.5$, respectively.

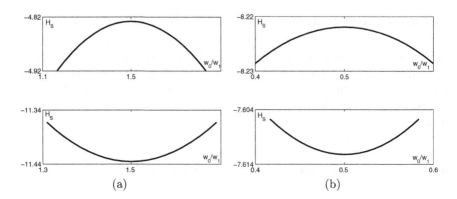

Fig. 3.14 $H_S(E)$ of the univariate tanh perceptron as a function of w_0/w_1: a) $(\mu_1,\ \sigma_1) = (3,\ 1)$ with $w_1 = 3$ at the top and $w_1 = 7$ at the bottom; b) $(\mu_1,\ \sigma_1) = (1,\ 1)$ with $w_1 = 10$ at the top and $w_1 = 15$ at the bottom.

Using the Nelder-Mead minimization algorithm [165, 132] it is possible to find the H_S-MEE solution $[w_1\ 0\ w_0]^T = [4.75\ 0\ -11.87]^T$, corresponding to the vertical line $x_1 = 2.5$, the optimal solution [219]. Fig. 3.15a shows H_S (represented by dot size) for a grid around the MEE solution: the central dot with minimum size. The Nelder-Mead algorithm was unable to find the optimal solution for the close-classes case. The reason is illustrated in Figs. 3.15b and 3.15c using a grid around the candidate solution. We encounter a minimum along w_1 and w_2, and a maximum along w_0. This means that, for close classes, if one only allows rotations of the optimal line the best MEE solution is in fact the vertical line. Note that by rotating the optimal line one is increasing the "disorder of the errors", with increased variance of the $f_{E|t}$ and increased errors for *both* classes; by Property 2 of Sect. 2.3.4 it is clear that the vertical line is an entropy minimizer. On the other hand, by shifting the vertical line along w_0 one finds a maximum because, when moving away of w_0^* one is increasingly assigning part of the errors to one of the classes and decreasing them for the other class; one then obtains the effect described by Property 1 of Sect. 2.3.4 and also seen in Example 3.2. We know already how empirical entropy solves this effect: using the fat estimate $\hat{f}_{fat}(e)$.

If empirical entropy is considered, all Gaussian-input problems can be solved with the MEE approach (with either $\hat{H}_S$ or $\hat{H}_{R_2}$), be they univariate or multivariate and with distant or close classes. The error rates obtained for sufficiently large n are very close to the theoretical ones (details provided in [219]). Examples for bivariate classes were already presented in the preceding section. Figure 3.16 shows the evolution of the weights for the $\mu_1 = [1\ 0]^T$ close-classes case, in an experiment where the perceptron was trained with $h = 1$ in $n = 500$ instances and tested in 5000 instances. The final solution is $\mathbf{w} = [0.54\ 0.05\ -0.23]^T$ corresponding to $\hat{P}_{ed} = 0.2880$ and $\hat{P}_{et} = 0.3119$ (we mentioned before that min $P_e = 0.3085$).

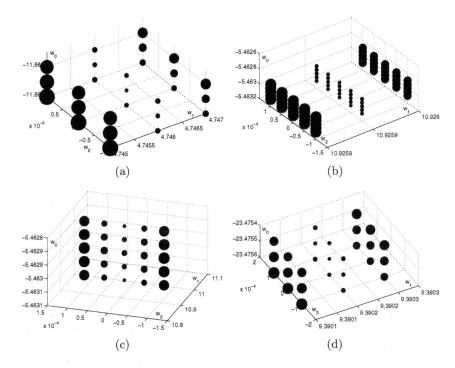

Fig. 3.15 H_S for bivariate Gaussian inputs, computed on a (w_1, w_2, w_0) grid around $\mathbf{w}^*$ (central dot). The dot size represents H_S value (larger size for higher H_S). (a) The distant-classes case, $\mu_1 = [5\ 0]^T$, with $\mathbf{w}^*$ the min H_S solution. (b) The close-classes case, $\mu_1 = [1\ 0]^T$, with $\mathbf{w}^*$ a minimizer for w_1 and w_2 and a maximizer for w_0. (c) A zoom of the central layer showing the maximum for w_0. (d) The same as (a) for the logistic activation function.

Similar results are obtained if the logistic activation function $\varphi(x) = 1/(1 + e^{-x})$ is considered. In this case

$$f_{E|t}(e) = \frac{\exp\left(-\frac{1}{2\sigma_t^2}\left(\ln\left(\frac{t-e}{1-t+e}\right) - m_t\right)^2\right)}{\sqrt{2\pi\sigma_t^2}(t-e)(1-t+e)}\mathbb{1}_{]-1+t,t[}(e). \tag{3.45}$$

For the same "distant classes" setting we find the H_S-MEE solution $[w_1\ 0\ w_0]^T = [9.3902\ 0\ -23.4755]^T$ (corresponding to the optimal solution $-w_0/w_1 = 2.5$) which is a local minimum as shown in Fig. 3.15(d). For the "close classes" setting, and in the same way as for the tanh activation function, a saddle point is found.

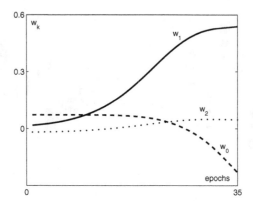

Fig. 3.16 Evolution of the weights in an experiment where the perceptron was trained to solve the $\mu_1 = [1\ 0]^T$ close-classes case, for bivariate Gaussian inputs.

3.3.2.2 Realistic Datasets

Comparison of theoretical and empirical MEE behaviors has to be restricted to two-dimensional problems, in order to have viable graphical representations and the Nelder-Mead algorithm running in reasonable time. One may, nonetheless, use dimensionally reduced real-world datasets. In what follows we consider the plane of the first two principal components (denoted $(x_1,\ x_2)$) of the original datasets when these have more than two features.

Since the true $(X_1,\ X_2)$ joint distributions are unknown, the true theoretical MEE solutions cannot be derived. One is still able, however, to derive theoretical MEE solutions of very closely resembling problems, proceeding in the following way: first, model the bivariate real-world PDFs by appropriate distributions, such that they achieve the same covariance matrices and with minimum L_1 distance of the marginal PDFs; next, apply to these modeled PDFs the procedure outlined in Sect. 3.1.1 (numerical simulation).

PDF modeling of the marginal class-conditional distributions is achieved by first obtaining from the data the Parzen window estimates, $\hat{f}_{X|t}$, using the optimal h_{IMSE} bandwidth. Next, one proceeds to adjust adequate known PDFs (namely, Gaussian, Gamma, and Weibull) by minimizing the L_1 distance between $\hat{f}_{X|t}$ and its model. The L_1 distance is preferable to other distance metrics (namely, L_2) by reasons described in [53]. Finally, for numerical computation of the theoretical MEE one generates a large number of points with the modelled class-conditional distributions and with the same estimated covariance matrix.

In the work [219] the datasets of Table 3.2 were analyzed. The datasets WDBC (30 features), Thyroid (5 features), and Wine (13 features) are from [13]; PB12, a dataset with 2 features, is from [110]. For the first three datasets the first two principal components were computed. These new datasets were named with a subscript '2' in Table 3.2 to distinguish them from the original ones. Next, the respective bivariate PDF models were obtained and an appropriate large number of instances generated maintaining the original class proportions. The total number of instances, also mentioned in Table 3.2, guaranteed IMSE < 0.01 of the estimated error PDFs. The empirical H_S-MEE solutions for the same datasets were computed, as well as the $\min P_e$ solutions with the Nelder-Mead algorithm.

For the multiclass datasets a sequential approach was followed, whereby the final classification was the result of successive dichotomies.

Table 3.2 shows the training set error rates obtained with both, theoretical and empirical, algorithms. They are in general close to the $\min P_e$ values (computed with the Nelder-Mead algorithm), the only exceptions being the theoretical MEE error rates for Thyroid$_2$ and PB12. Further details on these experiments are provided in the cited work [219].

Table 3.2 Error rates for the empirical and theoretical MEE algorithms, together with $\min Pe$ values, for four realistic datasets.

Dataset	No. classes	No. instances	Empirical MEE Error Rate	Theoretical MEE Error Rate	$\min P_e$
WDBC$_2$	2	2390	0.0824	0.0890	0.0808
Thyroid$_2$	3	2509	0.0367	0.0458	0.0375
Wine$_2$	3	5000	0.0553	0.0546	0.0526
PB12	4	6000	0.1072	0.1410	0.1067

The datasets with the decision borders achieved by the three algorithms are shown in Fig. 3.17. The decision borders are almost coincident, except the theoretical MEE borders for the Thyroid$_2$ and PB12 datasets.

3.3.3 The Arctangent Perceptron

Analytical expressions of theoretical EEs derived by the application of Theorem 3.2 can easily get quite involved, even for simple classifier settings. Usually, closed-form algebraic expressions of the entropies are simply impossible to obtain. A notable exception to this rule is Rényi's quadratic entropy of the arctangent perceptron with independent Gaussian inputs, which we study now. The arctangent perceptron (or arctan perceptron for short) is a perceptron whose activation function is the arctangent function.

Lemma 3.1. *The information potential of a two-class arctan perceptron (atan($\cdot$) activation function) fed with independent Gaussian inputs having mean μ_t and diagonal covariance matrix Σ_t, t denoting the class code, is*

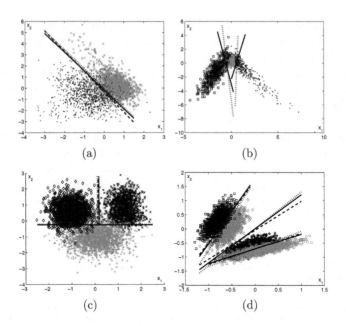

Fig. 3.17 Decision borders obtained with empirical MEE (dashed), theoretical MEE (dotted), and min P_e (solid) for WDBC$_2$ (a), Thyroid$_2$ (b), Wine$_2$ (c), and PB12 (d) datasets.

$$V_{R_2}(E) = \sum_t p_t^2 \left[\frac{1}{2\sqrt{\pi}\sigma_t}(1 + m_t^2) + \frac{\sigma_t}{4\sqrt{\pi}} \right] \qquad (3.46)$$

where p_t are the class priors, $m_t = \mathbf{w}^T\mu_t + w_0$, and $\sigma_t = \sqrt{\mathbf{w}^T\Sigma_t\mathbf{w}}$.

Proof. Let us denote the class-conditional sum of the independent Gaussian inputs by $U|t$. We know that:

$$U|t \in]-\infty, +\infty[\ \sim \ g(u; \mathbf{w}^T\mu_t + w_0, \sqrt{\mathbf{w}^T\Sigma_t\mathbf{w}}) = g(u; m_t, \sigma_t) \ . \qquad (3.47)$$

The sum of the inputs is submitted to the atan($\cdot$) sigmoid, i.e., the perceptron output is $Y = \varphi(U) = \mathrm{atan}(U) \in]-\pi/2, \ \pi/2[$. We could define $Y = \frac{2}{\pi}\mathrm{atan}(U) \in]-1, 1[$, but it turns out that this would unnecessarily complicate the following computations. We then also take $T = \{-\pi/2, \ \pi/2\}$; although different from what we are accustomed to (real-valued instead of integer-valued targets) it is perfectly legitimate to do so. When using t subscripts, we still assume $t \in \{-1, 1\}$ for notational simplicity (the class names are still "-1" and "1"), but what is really meant for the targets is $t\frac{\pi}{2}$.

We now determine $f_{Y|t}(y)$ using Theorem 3.2. We have:

$$u = \varphi^{-1}(y) = \tan(y); \ \varphi'(u) = \frac{1}{1 + u^2}; \ \varphi'(\varphi^{-1}(y)) = \frac{1}{1 + \tan^2(y)} \ . \qquad (3.48)$$

Therefore,

$$f_{Y|t}(y) = \frac{1}{\sqrt{2\pi}\sigma_t} \exp\left(-\frac{1}{2\sigma_t^2}(\tan(y) - m_t)^2\right)(1 + \tan^2(y)) . \qquad (3.49)$$

But from $f_{E|t}(e) = f_{Y|t}(t-e)$, follows that $f_{E|-1}(e) = f_{Y|-1}(-\frac{\pi}{2}-e) \in]-\pi, 0[$ and $f_{E|1}(e) = f_{Y|1}(\frac{\pi}{2} - e) \in]0, \pi[$. Thus:

$$f_{E|t}(e) = \frac{1}{\sqrt{2\pi}\sigma_t} \exp\left(-\frac{1}{2\sigma_t^2}(\tan(t\frac{\pi}{2} - e) - m_t)^2\right)(1 + \tan^2(t\frac{\pi}{2} - e)) = \qquad (3.50)$$

$$= \frac{1}{\sqrt{2\pi}\sigma_t} \exp\left(-\frac{1}{2\sigma_t^2}(\cot(e) - m_t)^2\right)(1 + \cot^2(e)) . \qquad (3.51)$$

Since the $f_{E|t}(e)$ are disjoint, $V_{R_2}(E) = p_1^2 V_{R_2}(E|1) + p_{-1}^2 V_{R_2}(E|-1)$, with

$$V_{R_2}(E|1) = \int_0^\pi f_{E|1}^2(e)de \quad \text{and} \quad V_{R_2}(E|-1) = \int_{-\pi}^0 f_{E|-1}^2(e)de . \qquad (3.52)$$

Let us perform the following change of variable: $z = \cot(e) - m_t$. We then have $\cot'(e)de = dz$, or $de = -\frac{dz}{1 + \cot^2(e)}$. Taking care of the limits of the integrals the potential can be expressed as

$$V_{R_2}(E|t) = \frac{1}{2\pi\sigma_t^2} \int_{+\infty}^{-\infty} e^{-z^2/\sigma_t^2} \frac{(1 + \cot^2(e))^2}{1 + \cot^2(e)}(-dz) = \qquad (3.53)$$

$$= \frac{1}{2\pi\sigma_t^2} \int_{-\infty}^{+\infty} e^{-z^2/\sigma_t^2}(1 + (z + m_t)^2)dz . \qquad (3.54)$$

On the other hand, with $\text{erf}(z) = \frac{2}{\sqrt{\pi}} \int_0^z e^{-t^2} dt$ ($\text{erf}(-\infty) = -1$, $\text{erf}(+\infty) = 1$), we have:

$$\frac{1}{2\pi\sigma^2} \int_{-\infty}^{+\infty} e^{-z^2/\sigma^2} dz = \frac{1}{4\sqrt{\pi}\sigma}\left[\text{erf}\left(\frac{z}{\sigma}\right)\right]_{-\infty}^{+\infty} = \frac{1}{2\sqrt{\pi}\sigma} ; \qquad (3.55)$$

$$\frac{1}{2\pi\sigma^2} \int_{-\infty}^{+\infty} e^{-z^2/\sigma^2}(z + m)^2 dz =$$

$$= \frac{1}{2\pi\sigma^2}\left[-\frac{\sigma^2 z}{2}e^{-z^2/\sigma^2} + \frac{\sigma^3\sqrt{\pi}}{4}\text{erf}\left(\frac{z}{\sigma}\right) - m\sigma^2 e^{-z^2/\sigma^2} + \right.$$

$$\left. + \frac{m^2\sqrt{\pi}\sigma}{2}\text{erf}\left(\frac{z}{\sigma}\right)\right]_{-\infty}^{+\infty} = \frac{1}{2\pi\sigma^2}\left[\frac{\sigma^3\sqrt{\pi}}{2} + m^2\sqrt{\pi}\sigma\right] = \frac{\sigma}{4\sqrt{\pi}} + \frac{m^2}{2\sqrt{\pi}\sigma} . \qquad (3.56)$$

Substituting (3.55) and (3.56) in (3.53) we finally obtain the expression (3.46) of $V_{R_2}(E)$.   □

Remarks:

1. Note that the supports of $f_{E|-1}(e)$ and $f_{E|1}(e)$ are disjoint; respectively, $]-\pi, 0[$ and $]0, \pi[$. This is a consequence of using a sigmoidal activation function, and contrasts to what happened with the linear discriminant.
2. The expression of $V_{R_2}(E)$ has terms m_t^2 containing the bias w_0. Therefore, and also contrasting with the linear discriminant, the MEE algorithm is now always able to adjust the bias of the decision borders. The same remark applies to EE for formulas (3.44) and (3.45).

It is a well-known fact that, for two-class problems with Gaussian inputs having equal covariance matrix Σ, the Bayes linear decision function for equal priors [182, 76] is

$$d(\mathbf{x}) = [\mathbf{x} - \frac{1}{2}(\mu_{-1} + \mu_1)]^T \Sigma^{-1}(\mu_{-1} - \mu_1) ; \qquad (3.57)$$

in other words, the linear discriminant $d(\mathbf{x}) = 0$ passes through the point lying half-way of the means — $\frac{1}{2}(\mu_{-1} + \mu_1)$ — and is orthogonal to $\Sigma^{-1}(\mu_{-1} - \mu_1)$. We now analyze how the arctan perceptron behaves relatively to this issue in the next theorem.

Theorem 3.3. *The Bayes linear discriminant for equal-prior two-class problems, with Gaussian inputs having the same covariance Σ, is a critical point of Rényi's quadratic entropy of the arctan perceptron.*

Proof. We first note that, given a two-class problem with Gaussian inputs S having the same covariance Σ, one can always transform it into an equivalent problem with covariance $\mathbf{I}$, by applying the *whitening transformation* to the inputs (see e.g. [76]): $\mathbf{x} = (\Phi \Lambda^{-1/2})^T \mathbf{s}$, where $\mathbf{s}$ is a (vector) instance of the multi-dimensional r.v. S, Φ and Λ are the eigenvector and eigenvalue matrices of Σ, respectively, and $\mathbf{x}$ is the corresponding instance of the whitened multi-dimensional r.v. X. Since the whitening transformation is a linear transformation, the position of the critical points of a continuous function of S, like the one implemented by the arctan perceptron, is also linearly transformed by $(\Phi \Lambda^{-1/2})^T$. For the whitened inputs the Bayes discriminant is orthogonal to the vector linking the means $\mu_{-1} - \mu_1$.

We then apply the previous Lemma to the whitened inputs problem, and proceed to studying the critical points of $V(\mathbf{w}, w_0) \equiv 8\sqrt{\pi} V_{R_2}(E; \mathbf{w}, w_0)$:

$$V \equiv V(\mathbf{w}, w_0) = \sum_t \left[\frac{1}{\sigma}(1 + m_t^2) + \frac{\sigma}{2} \right] = \frac{1}{\sigma}(1 + m_{-1}^2 + m_1^2) + \sigma , \quad (3.58)$$

with $\sigma = \sqrt{\mathbf{w}^T \mathbf{I} \mathbf{w}} = \|\mathbf{w}\|$. Let us take the partial derivatives. For the bias term:

$$\frac{\partial V}{\partial w_0} = \frac{2}{\sigma}(m_{-1} + m_1) = 0 \;\Rightarrow\; \begin{cases} m_1 = -m_{-1} = \frac{1}{2}\mathbf{w}^T(\boldsymbol{\mu}_1 - \boldsymbol{\mu}_{-1}) \\ w_0 = -\frac{1}{2}\mathbf{w}^T(\boldsymbol{\mu}_1 + \boldsymbol{\mu}_{-1}) \end{cases} . \quad (3.59)$$

The critical points of V do indeed constrain the discriminant to pass through the point lying half-way of the means. Let us now rotate the coordinate system of X such that one of the coordinates, say k, passes through the mean vectors $\boldsymbol{\mu}_t$, and shift its origin to $\frac{1}{2}(\boldsymbol{\mu}_1 + \boldsymbol{\mu}_{-1}) = 0$; therefore, setting $w_0 = 0$. This rotation and shift amount to an additional linear transformation of the inputs. Since in the new coordinate system, $\mu_{t,i} = 0$ and $\mu_{1,k} = -\mu_{-1,k} = \mu$, we then have:

$$V = \frac{2}{\sigma}(1 + w_k^2 \mu^2) + \sigma . \quad (3.60)$$

For the partial derivatives relative to the weights w_i, we distinguish two cases:

1. $i \neq k$:

$$\frac{\partial V}{\partial w_i} = 0 \;\Rightarrow\; \frac{w_i}{\sigma^3}(\sigma^2 - 2 - 2w_k^2\mu^2) = 0 . \quad (3.61)$$

For $\sigma \neq 0$, the solutions are $w_i = 0$, $i \neq k$, and $\sigma^2 = 2 + 2w_k^2\mu^2$.

2. $i = k$:

$$\frac{\partial V}{\partial w_k} = 0 \;\Rightarrow\; \frac{w_k}{\sigma^3}(\sigma^2(1 + 4\mu^2) - 2 - 2w_k^2\mu^2) = 0 . \quad (3.62)$$

For $\sigma \neq 0$, the solutions are $w_k = 0$ and $\sigma^2(1 + 4\mu^2) = 2 + 2w_k^2\mu^2$.

Combining the solutions of 1. and 2. the critical points, for $\sigma \neq 0$ and $\mu \neq 0$, are readily seen to be:

$$w_{i \neq k} = 0, \quad w_k = \pm\sqrt{\frac{2}{1 + 2\mu^2}}; \quad (3.63)$$

$$w_k = 0, \quad \sum_{i \neq k} w_i^2 = 2 . \quad (3.64)$$

The solution $w_{i \neq k} = 0$ (3.63) corresponds to a hiperplane orthogonal to the vector linking the means, i.e., to the Bayes solution for whitened inputs. Once this hyperplane has been found, we may revert to the original system of coordinates and apply the inverse whitening transformation $(\mathbf{\Lambda}^{1/2}\mathbf{\Phi}^T)^T\mathbf{x}$. We then get expression (3.57) because $\mathbf{x}^T\mathbf{x}$ transforms into $\mathbf{s}^T\mathbf{\Phi}\mathbf{\Lambda}^{-1/2}\mathbf{\Lambda}^{-1/2}\mathbf{\Phi}^T\mathbf{s} = \mathbf{s}^T\mathbf{\Sigma}^{-1}\mathbf{s}$ since $\mathbf{\Phi}^T = \mathbf{\Phi}^{-1}$ and $\mathbf{\Sigma}^{-1} = (\mathbf{\Phi}\mathbf{\Lambda}\mathbf{\Phi}^T)^{-1} = (\mathbf{\Phi}^T)^T\mathbf{\Lambda}^{-1}\mathbf{\Phi}^T$. $\quad\square$

The type of the critical points (3.63) and (3.64) is determined by the eigenvalues of the Hessian. First note that in (3.60), V does not depend on the bias. For the weights, setting $A = 1 + w_k^2\mu^2$ and $B = 1 + 4\mu^2$:

1. $i \neq k$:

$$\frac{\partial^2 V}{\partial w_i^2} = \frac{1}{\sigma^3} \left[\left(\frac{6w_i^2}{\sigma^2} - 2 \right) A + \sigma^2 - w_i^2 \right] \Rightarrow$$

$$\Rightarrow \begin{cases} \left. \dfrac{\partial^2 V}{\partial w_i^2} \right|_{w_{i \neq k} = 0} = \dfrac{-8\mu^2 \sqrt{1+2\mu^2}}{2\sqrt{2}} < 0 \\[3mm] \left. \dfrac{\partial^2 V}{\partial w_i^2} \right|_{w_k = 0} = \dfrac{w_i^2}{\sqrt{2}} > 0 \end{cases} \tag{3.65}$$

$$\frac{\partial^2 V}{\partial w_k \partial w_i} = \frac{w_i w_k}{\sigma^3} \left[\frac{6A}{\sigma^2} - B \right] \Rightarrow \left. \frac{\partial^2 V}{\partial w_k \partial w_i} \right|_{w_{i \neq k} = 0} = \left. \frac{\partial^2 V}{\partial w_k \partial w_i} \right|_{w_k = 0} = 0 \tag{3.66}$$

2. $i = k$:

$$\frac{\partial^2 V}{\partial w_k^2} = \frac{A}{\sigma^3} \left(\frac{6w_k^2}{\sigma^2} - 2 \right) - \frac{4w_k^2 \mu^2}{\sigma^2} + \frac{B}{\sigma} \left(1 - \frac{w_k^2}{\sigma^2} \right) \Rightarrow$$

$$\Rightarrow \begin{cases} \left. \dfrac{\partial^2 V}{\partial w_k^2} \right|_{w_{i \neq k} = 0} = \dfrac{4}{\sigma^3} > 0 \\[3mm] \left. \dfrac{\partial^2 V}{\partial w_k^2} \right|_{w_k = 0} = \dfrac{4\mu^2}{\sqrt{2}} \end{cases} \tag{3.67}$$

3. $j \neq i$ and $i, j \neq k$:

$$\frac{\partial^2 V}{\partial w_j \partial w_i} = \frac{w_i w_j}{\sigma^3} \left[\frac{6A}{\sigma^2} - 1 \right] \Rightarrow \begin{cases} \left. \dfrac{\partial^2 V}{\partial w_j \partial w_i} \right|_{w_{i \neq k} = 0} = 0 \\[3mm] \left. \dfrac{\partial^2 V}{\partial w_j \partial w_i} \right|_{w_k = 0} = \dfrac{w_i w_j}{\sqrt{2}} \end{cases} . \tag{3.68}$$

Therefore, the Hessian for the $w_{i \neq k} = 0$ critical points is a diagonal matrix whose elements are precisely the eigenvalues. Thus, we get saddle points. In what concerns the $w_k = 0$ critical points they are either minima or saddle points.

Note that the $\mu = 0$ setting would produce an infinity of $\sum_i w_i^2 = 2$ critical points, but this is an uninteresting degenerate configuration with no distinct classes. The $\sigma = 0$ setting is discussed in the following example.

Example 3.9. We apply the results of Theorem 3.3 to the two-dimensional case. For whitened Gaussian inputs with means lying on x_1, symmetric about x_2 and with $\mu = 1$, we have two critical points of $V_{R_2}(w_1, w_2)$ located at $w_0 = 0$ and at the following weight vectors: $\left[\pm \sqrt{2/3} \ 0 \right]^T$, $\left[0 \ \pm \sqrt{2} \right]^T$.

Figure 3.18 shows a segment of the $V_{R_2}(w_1, w_2)$ surface with the two critical points for positive weights signalled by vertical lines. Figure 3.19 shows the error PDFs at these two critical points. The $[0 \ \pm \sqrt{2}]^T$ critical points are minima (confirming the above results) and correspond to a line passing through the means; one indeed expects the entropy to be a maximum for this setting, and the information potential a minimum.

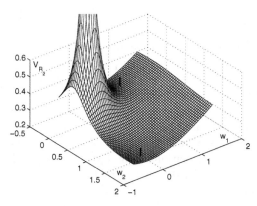

Fig. 3.18 A segment of the Rényi's information potential for the two-dimensional arctan perceptron ($w_0 = 0$, $\mu = 1$) with vertical lines signaling the critical points.

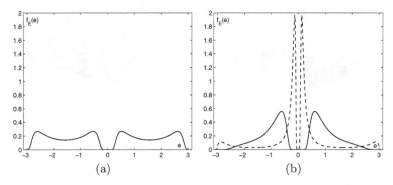

Fig. 3.19 Error PDFs of the arctan perceptron at three $[w_1 \; w_2]^T$ points (Example 3.9): (a) critical point $[0 \; \sqrt{2}]^T$ with $V_{R_2} = 0.1995$; (b) critical point $[\sqrt{2/3} \; 0]^T$ with $V_{R_2} = 0.3455$ (solid line) and at $[4 \; 0]^T$ with $V_{R_2} = 0.8815$ (dashed line).

Figure 3.18 also shows an infinite maximum at $[0 \; 0]^T$ corresponding to $\sigma = 0$. This zero-weights maximum is of no practical consequences. For instance, when considering the application of gradient ascent to $V_{R_2}(w_1, w_2)$, at a certain point the gradient is so high (error PDFs close to Dirac-δ functions at $\pm\frac{\pi}{2}$) that it will exceed machine range and cause an overshoot to some other $[w_1 \; w_2]^T$ point.

We are then left with the only interesting critical point, with $w_2 = 0$, a saddle point as predicted by expressions (3.65) to (3.68). From this point onwards, progressing towards higher w_1 values along the $w_2 = 0$ direction, one obtains increasing $V_{R_2}(w_1, w_2)$ values corresponding to error PDFs approaching two Dirac-δ functions at the origin. Figure 3.19 shows the error PDF at $[4 \; 0]^T$, illustrating what happens for growing w_1. $\qquad\square$

If one convolves the error PDFs with a Gaussian kernel, G_h, something similar to Example 3.2 does happen. The theoretical infinite maximum at the origin is removed, due to kernel smoothing, and a potential hill along $w_2 = 0$ emerges. Figure 3.20a shows the potential surface corresponding to $V_{R_2}(G_{1.5} \otimes f_E; w_1, w_2)$ for the same $w_0 = 0$ setting as in Fig. 3.18.

A potential hill is also evident when computing empirical potential with fat estimation of the error samples. Figure 3.20b shows the empirical potential surface, $\hat{V}_{R_2}(\hat{f}_E; w_1, w_2)$, based on fat estimation of $\hat{f}_E$ for an experimental run with 100 error samples and bandwidth $h = 1$. Gradient ascent in this surface provides a good estimation of the optimum hyperplane, converging to a solution around $\mathbf{w}^* = [4\ 0]^T$, with practically zero gradient thereafter along the $w_2 = 0$ crest.

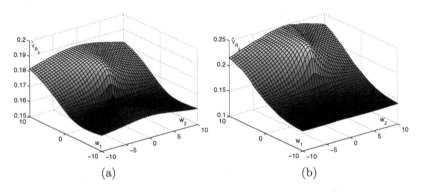

(a)   (b)

Fig. 3.20 Surfaces of the information potential for the arctan perceptron: a) $V_{R_2}(G_{1.5} \otimes f_E; w_1, w_2)$; b) $\hat{V}_{R_2}(\hat{f}_E; w_1, w_2)$ based on fat estimation of $\hat{f}_E$ for 100 error samples (kernel bandwidth $h = 1$).

When applied to Gaussian data with equal class covariance, the arctan perceptron with information potential risk (or, equivalently, with quadratic Rényi *EE* risk) behaves adequately, in terms of attaining min P_e solutions. Table 3.3 shows one-experiment results obtained for the unit covariance case we have been discussing, and with gradient ascent applied to the information potential in 65 epochs. The data consisted of 2000 instances per class and the kernel bandwidth was $h = 1$; optimal PDF estimation would require $h = 0.31$, therefore fat estimation is indeed being used. The weight values shown in Table 3.3 are normalized to $w_1 = 1$. The error rates at the end of the 65 epochs, together with the Bayes optimal min P_e values, are also shown. The arctan perceptron is clearly producing solutions close to the optimal ones.

Table 3.4 shows one-experiment results when the covariance is

$$\Sigma = \begin{bmatrix} 1 & 0.5 \\ 0.5 & 2 \end{bmatrix}. \tag{3.69}$$

In this case the optimal w_1-normalized weight vector is $[1 \ -0.25]^T$. The Bayes optimal $\min P_e$ values were computed with the formulas provided by [191]. The experimental error rate results obtained with 2000 instances per class and 65 epochs training are again in close agreement with the theoretical values.

Similar experiments found the arctan perceptron with Shannon EE risk having an analogous behavior to that of the arctan perceptron with quadratic Rényi EE risk. The arctan perceptrons also behave properly when applied to real-world datasets; performance figures of MLPs using these perceptrons (and of other types as well) applied to classification tasks of real-world datasets are presented later, in Sect. 6.1.3.

Table 3.3 Results of the H_{R_2} arctan perceptron applied to Gaussian data with unit covariance (see text).

μ	w_1	w_2	w_0	$\hat{P}_e$	$\min P_e$
0.2	1	-0.0049	-0.1639	0.4323	0.4207
0.5	1	-0.0074	0.0043	0.2990	0.3085
1	1	-0.0433	-0.0231	0.1570	0.1597
1.5	1	-0.0100	0.0029	0.0673	0.0668
2	1	-0.0408	-0.0180	0.0220	0.0228

Table 3.4 Results of the H_{R_2} arctan perceptron applied to Gaussian data with covariance (3.69) (see text).

μ	w_1	w_2	w_0	$\hat{P}_e$	$\min P_e$
0.2	1	-0.2592	-0.0165	0.4000	0.4253
0.5	1	-0.2629	0.0642	0.3040	0.2965
1	1	-0.2387	-0.0853	0.1470	0.1425
1.5	1	-0.2445	0.0081	0.0517	0.0544
2	1	-0.2117	-0.0155	0.0143	0.0163

We conclude the present section with the presentation of Table 3.5, providing a synoptic comparison of perceptrons with the three activation functions that we analyzed, regarding the relation of $\min P_e$ to critical points of the error entropy, and this for both the theoretical and empirical settings. For the empirical setting, besides the experiments described in this and preceding sections, we also took into consideration a large amount of experimental evidence derived from the application of MLPs to solving real-world classification tasks, part of which are reported in Chap. 6.

Table 3.5 Synoptic comparison of perceptrons according to the activation function.

Sigmoid a.f.	Shannon's EE H_S Gaussian inputs	$\hat{H}_S$	Rényi's Quadratic EE H_{R_2} Gaussian inputs	$\hat{H}_{R_2}$
Tanh	min P_e at min EE for w.s.c.[a]; at max EE otherwise.	min $\hat{P}_e$ at min EE, p.a.[b]	Analysis not available.	min $\hat{P}_e$ at min EE, p.a.[b]
Logistic	min P_e at min EE for w.s.c.[a]; at max EE otherwise.	min $\hat{P}_e$ at min EE, p.a.[b]	Analysis not available.	min $\hat{P}_e$ at min EE, p.a.[b]
Arctan	Analysis not available.	min $\hat{P}_e$ at min EE, p.a.[b]	min P_e at a saddle point of EE.	min $\hat{P}_e$ at min EE, p.a.[b]

a. w.s.c. stands for "well separated classes", usually with distance between the means exceeding 1.5 times the standard deviation.
b. p.a. stands for "presumably always".

3.4 The Hypersphere Neuron

The hypersphere neuron implements the following classifier function family:

$$\mathcal{Z}_W = \left\{ \theta(\varphi(\|\mathbf{x} - \mathbf{w}\|^2 - w_0^2); \mathbf{w} \in W \subset \mathbb{R}^d, r \in \mathbb{R} \right\} . \qquad (3.70)$$

The argument of the activation function $\varphi(\cdot)$ defines a hyperparaboloid in d-dimensional space. Setting $\varphi(\|\mathbf{x} - \mathbf{w}\|^2 - w_0^2) = a$ a hyperspherical decision border is obtained.

The hypersphere classifier was first used as a building block in a more sophisticated type of classifier, the so-called compound classifier [18, 19]. In more recent times, it has attracted the attention of many researchers of the Pattern Recognition area due to the ease of use and efficiency revealed by networks of hyperspheres [119, 16, 17, 236]. It has also inspired some new approaches on RBF NNs [58].

The training algorithm for the hypersphere neuron, using gradient descent, follows exactly the same steps mentioned in Sect. 3.1.2. In the following experiments we only consider the hypersphere classifier in bivariate space; the decision border is then a circle.

3.4.1 Motivational Examples

Example 3.10. In this example we use two Gaussian distributed class-conditional PDFs, $g(\mathbf{x}; \mu_t, \Sigma_t)$, with

$$\mu_{-1} = [2\ 0]^T, \ \mu_1 = [0\ 0]^T, \text{ and } \Sigma_{-1} = \Sigma_1 = \mathbf{I}.$$

We now present an experiment with 200-instance training and test datasets (100 instances per class) and apply the (Shannon) MEE gradient descent algorithm with $h = 1$ (fat estimation) and $\eta = 0.001$. Fig. 3.21 shows the evolution of the decision border and the error PDF (computed only at the e_i values) at successive epochs. Note the evolution towards a monomodal PDF, as the position of the circle center is being adjusted. Fig. 3.22 shows the converged solution. In the final epochs the radius of the circle undergoes the main adjustments.   □

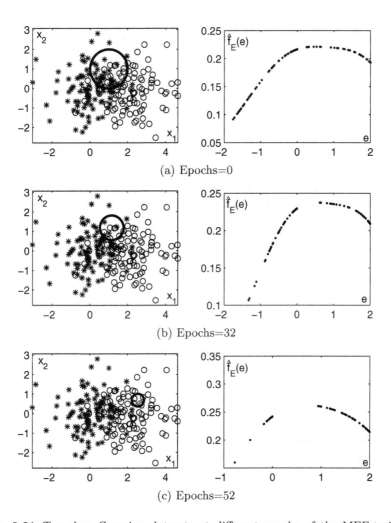

Fig. 3.21 Two-class Gaussian datasets at different epochs of the MEE-trained circle. The left graphs show the datasets with the circular decision border (solid line). The right graphs show the error PDF in the $E = [-2, 2]$ support.

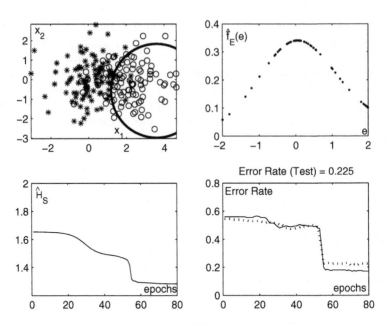

Fig. 3.22 The final converged solution of Example 3.10. The downside graphs of the Shannon entropy and the error rate (solid line for the training set and dotted line for the test set) are in terms of the no. of epochs.

Example 3.11. In this example we consider bivariate datasets for which the min P_e solution is easy to find. We call them *ring* datasets. They consist of a Gaussian dataset with $\mu_{-1} = [0 \ \ 0]^T$ and $\Sigma_{-1} = \lambda I$ for class -1, and a dataset with uniform distribution in a circular ring (annulus) centered at $\mu_1 = [0 \ \ 0]^T$, defined as

$$f(\mathbf{x}) = \begin{cases} \frac{1}{\pi(r_2^2 - r_1^2)} & r_1^2 \leq \|\mathbf{x} - \mu\|^2 \leq r_2^2 \\ 0 & \text{otherwise} \end{cases}, \tag{3.71}$$

for class 1.

The min P_e value for this kind of datasets is easy to compute. In fact, by symmetry, the optimal decision border is a circle centered at the origin and whose radius we call a. For priors p and $q = 1 - p$, class 1 contribution to min P_e is then $p\frac{a^2 - r_1^2}{r_2^2 - r_1^2}$; class -1 contribution is $q \exp(-a^2/(2\lambda))$ (formulas for the probability of a Gaussian distribution in an elliptical area can be found in [144]).

We now present the results of an experiment with equal priors, $r_1 = 1$, $r_2 = 2.4$, and $\lambda = 0.3$, for which min $P_e = 0.0885$ ($a_{opt} = 1.1147$). The datasets in this experiment have 300 instances per class and the H_S-MEE-training of the circle is carried out during 60 epochs with $h = 1$ (fat estimation) and $\eta = 0.001$.

Fig. 3.23 shows the evolution of the decision border and the error PDF (computed only at the e_i values) at successive epochs. Fig. 3.24 shows the converged solution. In the final epochs the radius of the circle undergoes the main adjustments converging to a value close to a_{opt}.

Finally, Fig. 3.25 shows the learning curves for this experimental setting, evidencing a consistent learning of the MEE-trained circle.   □

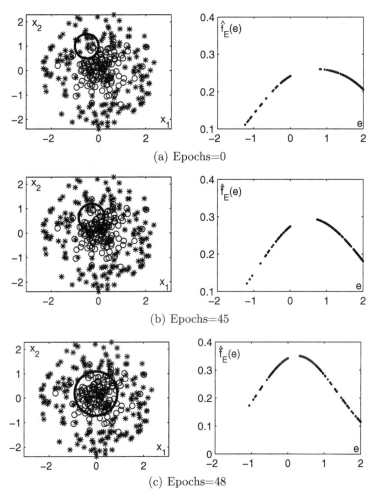

(a) Epochs=0

(b) Epochs=45

(c) Epochs=48

Fig. 3.23 The ring dataset of Example 3.11 at different epochs of the MEE-trained circle. The left graphs show the dataset with the circular decision border (solid line). The right graphs show the error PDF in the $E = [-2, 2]$ support.

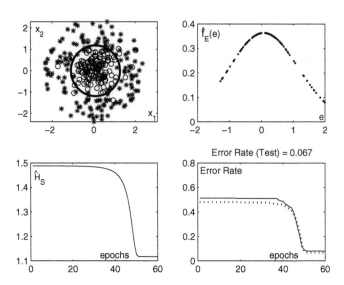

Fig. 3.24 The final converged solution of Example 3.11. The downside graphs of the Shannon entropy and the error rate (solid line for the training set and dotted line for the test set) are in terms of the no. of epochs.

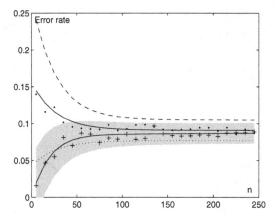

Fig. 3.25 Learning curves for the ring dataset of Example 3.11. The learning curves were obtained by exponential fits to the $\overline{P}_{ed}(n)$ (denoted '+') and $\overline{P}_{et}(n)$ (denoted '.') values. The shadowed region represents $\overline{P}_{ed} \pm s(P_{ed})$; the dashed lines represent $\overline{P}_{et} \pm s(P_{et})$.

3.4.2 Theoretical and Empirical MEE in Realistic Datasets

We follow the same procedure as in Sect. 3.3.2.2, applying the learning algorithm of the hypersphere neuron (in fact, the circle neuron) with

Table 3.6 Error rates for the empirical and theoretical MEE algorithms, together with min P_e values, for three datasets.

Dataset	No. classes	No. instances	Empirical MEE Error Rate	Theoretical MEE Error Rate	min P_e
Wine$_2$	3	5000	0.0415	0.0512	0.0402
Thyroid$_2$	3	2509	0.0494	0.2152	0.0359
Ionosphere$_2$	2	5778	0.1888	-	0.1865

theoretical and empirical SEE. We present the results reported in [219] for the datasets Wine$_2$, Thyroid$_2$, and Ionosphere$_2$, i.e., the versions of the original Wine, Thyroid, and Ionosphere datasets made of their first two principal components (the original Ionosphere dataset is from [13]). The aim in these experiments is simply to discriminate one class from the other ones.

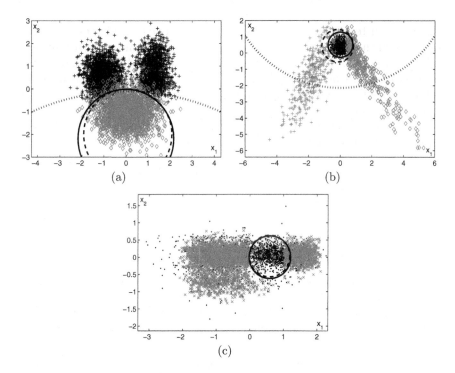

Fig. 3.26 Decision borders obtained with empirical MEE (dashed), theoretical MEE (dotted), and min P_e (solid) for Wine$_2$ (a), Thyroid$_2$ (b), and Ionosphere$_2$ (c) datasets.

Table 3.6 shows the training set error rates obtained with both, theoretical and empirical, algorithms. They are in general close to the min P_e values (obtained with the Nelder-Mead algorithm), the only exceptions being the theoretical MEE error rates for Thyroid$_2$ and Ionosphere$_2$: in this last case, characterized by a large distribution overlap, the theoretical MEE algorithm assigned all errors to one of the classes and was unable to converge to a more balanced solution. Further details on these experiments are provided in [219].

Figure 3.4.2 shows the datasets with the decision borders achieved by the hypersphere (circle) neuron. It is clear that the empirical MEE always finds a good solution close to the min Pe solution. The theoretical MEE algorithm exhibits the difficulties that were already seen in the preceding section and correspond to getting stuck in a local minimum solution which is inefficient. This effect is basically the one that we had already found in the simple Example 3.2. Further details on the theoretical MEE solutions are presented in [219].

3.5 The Data Splitter

The data-splitter is one of the simplest classifiers one can think of; its task is to find an optimal split point in univariate data. The continuous output data splitter can be viewed as a perceptron with only one adjustable parameter, the bias w_0, with output expressed by $y = \varphi(x - w_0)$. We then have, by virtue of (2.22), that the perceptron is trained to find a split point in the error distribution.

Assuming the perceptron uses the $\tanh(\cdot)$ activation function, we apply Theorem 3.2 noting that:

$$\varphi'(x) = 1 - \tanh^2(x - w_0) \neq 0 \ \forall x \tag{3.72}$$

$$\varphi^{-1}(y) = w_0 + \mathrm{atanh}(y) \ , \tag{3.73}$$

$$\varphi'(\varphi^{-1}(y)) = 1 - y^2 \ . \tag{3.74}$$

For the $\tanh(\cdot)$ activation function, since $y \in]-1, 1[$, we use the target value set $T = \{-1, 1\}$.

Using the above formulas let us see how the data splitter theoretically performs for the simple case of uniform class-conditional inputs:

$$f_{X|-1}(x) = \frac{1}{b - a}\mathbb{1}_{[a,b]}(x), \ f_{X|1}(x) = \frac{1}{d - c}\mathbb{1}_{[c,d]}(x) \tag{3.75}$$

with $a < c \leq b < d$.

The error class-conditional PDFs can be obtained using Theorem 3.2, and both Shannon's and Rényi's quadratic entropies can be thereafter derived [212]. The expression of the quadratic Rényi information potential is

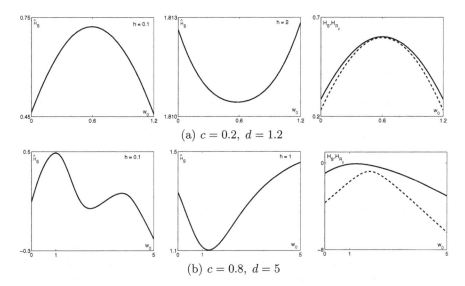

(a) $c = 0.2$, $d = 1.2$

(b) $c = 0.8$, $d = 5$

Fig. 3.27 The left and middle graphs show $\hat{H}_S$ as a function of w_0 for different values of h and w_1 support $[c, d]$. The right-most graphs are the corresponding theoretical curves of Shannon (solid) and Rényi's (dotted) entropies.

$$V_{R_2} = -\frac{q^2}{4} \left[\frac{2 + e(e+2)\ln\left(\frac{|e|}{e+2}\right) + 2e}{(b-a)^2(e+2)e} \right]_{-1-\tanh(b-w_0)}^{-1-\tanh(a-w_0)} +$$

$$+ \frac{p^2}{4} \left[\frac{2 + e(e-2)\ln\left(\frac{e}{|e-2|}\right) - 2e}{(d-c)^2(e-2)e} \right]_{-1-\tanh(d-w_0}^{-1-\tanh(c-w_0)} . \tag{3.76}$$

The expression of the Shannon entropy is even more intricate [212]. For both entropies it is possible to show that neither of them has a minimum at the optimal split point (the min P_e point); as also shown in the cited work, for class-conditionals with equal-length supports, they have a maximum.

Figure 3.27 compares the behaviors of theoretical and empirical (2000 instances per class) Shannon entropy curves as functions of the split parameter w_0, varying the class-conditional configuration. In all cases we assume equal priors ($p = q = 1/2$) and fix w.l.o.g. $[a, b] = [0, 1]$.

In the top row of Fig. 3.27 the class-conditionals have equal-length support, which means that the optimal solution is any point of the overlapped region $[c, 1]$. If the kernel bandwidth h is too small, $\hat{H}_S$ exhibits a maximum at the optimal split point, just as H_S.

Above a sufficiently large h (and increasingly larger with increasing overlap), the Shannon MEE solution is indeed the optimal split point.

The bottom row of Fig. 3.27 illustrates the unequal class error probabilities case (due to increased ω_1 support) with optimal split point $w_0^* = 1$. Both the theoretical and empirical curves fail to find w_0^*. However, the empirical minimum occurs in a close neighborhood of w_0^*.

For Gaussian class-conditional distributions of the perceptron input there are no closed-form expressions of the entropies. Using numerical computation of the integrals, similar conclusions can be drawn (theoretical entropy maximum at the min P_e point; empirical entropy minimum close to the min P_e point).

3.6 Kernel Smoothing Revisited

The present section analyzes in greater detail the influence of kernel smoothing in the attainment of MEE solutions for continuous error classifiers. The analysis follows the exposition given in [212].

Consider the data splitter with Gaussian inputs as in the previous section. Figure 3.28 illustrates the influence of kernel smoothing in the error PDF estimation. The figure shows the theoretical and empirical PDFs for two locations of the split point: off-optimal (3.28a) and optimal (3.28b). Note the smoothing imposed by the kernel estimate: an increased h implies a smoother estimate with greater impact near the origin.

The bottom graphs of Fig. 3.28a and 3.28b are illustrative of why a theoretical maximum can change to an empirical minimum. The error PDF in (a) is almost uniform for class ω_{-1}, implying a high value of $H_{S|-1}$; however, the error PDF for class ω_1 is highly concentrated implying a very low $H_{S|1}$; $f_{E|1}$ is clearly more concentrated than its left counterpart. Property 3 (Sect. 2.3.4) and formula (C.3) give then the plausible justification of why the overall H_S turns out to be smaller for the off-optimal than for the optimal split point.

With the error PDF estimated with a sufficiently high value of h we get the sort of curves showed with dotted line at the bottom graphs of (a) and (b). Kernel smoothing "couples" the class-conditional components of the error PDF, which is then seen as a "whole", ignoring relation (C.3); the density for the nonoptimal split has now a long tail, whereas the density of the optimal split is clearly more concentrated at the origin. As a consequence a minimum of the entropy is obtained at the optimal split point. A similar maximum-to-minimum entropy flip due to kernel smoothing is observed in other classifiers, namely those discussed in the present chapter.

We now analyze the theoretical behavior of the kernel smoothing effect on two distinct PDFs that resemble the ones portrayed in Fig. 3.28. One of them, $f_1(x)$, corresponds to the off-optimal error PDF with a large tail for one class and a fast-decaying trend for the other class, modeled as

$$f_1(x; \lambda) = \frac{1}{2} u(x; -1, 0) + \frac{1}{2} e_+(x; \lambda),$$

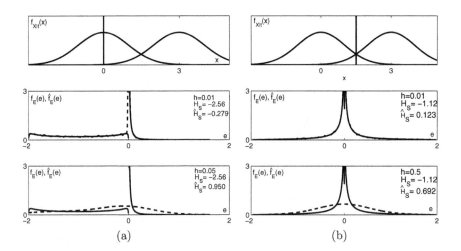

Fig. 3.28 Kernel smoothing effect for two locations of the split point: off-optimal (a) and optimal (b). The top graphs are the Gaussian input PDFs. The middle and bottom graphs show the theoretical (solid line) and smoothed (dotted line) error PDFs for two different values of h.

where $e_+(x; \lambda)$ is the exponential PDF with parameter λ, decaying for $x \geq 0$. The other PDF, $f_2(x)$, corresponds to the optimal error PDF modeled as

$$f_2(x; a) = \frac{1}{2}e_+(x; a) + \frac{1}{2}e_-(x; a),$$

where $e_-(x; a)$ is the symmetrical of $e_+(x; a)$.

Simple calculations of the Rényi entropies (for Shannon entropy, the problem is intractable) show that the respective entropies H_1 and H_2 are such that $H_1 < H_2$ (i.e., a maximum as in Fig. 3.28) for $\lambda > 2(a - 1)$. We now proceed to convolve these PDFs with a Gaussian $G_h(\cdot)$. The resulting PDFs are

$$G_h \otimes f_1(x) = \frac{1}{2}\left[\Phi\left(\frac{x+1}{h}\right) - \Phi\left(\frac{x}{h}\right)\right] + \frac{\lambda}{2}e^{\frac{\lambda^2 h^2}{2} - \lambda x}\Phi\left(\frac{x - \lambda h^2}{h}\right)$$

$$G_h \otimes f_2(x) = \frac{a}{2}e^{\frac{a^2 h^2}{2}}\left[e^{ax}\left[1 - \Phi\left(\frac{x + ah^2}{h}\right)\right] + e^{-ax}\left[1 - \Phi\left(\frac{x - ah^2}{h}\right)\right]\right]$$

where $\Phi(x)$ denotes $\int_{-\infty}^{x} g(u; 0, 1)du$, the standard Gaussian CDF.

Using these formulas and setting $\lambda = 2(a - 1) + 1$ in order to have an entropy maximum for the original f_2 PDF, one may always find a sufficiently large h such that the kernel smearing out of the tail of f_1 will turn the entropy of f_2 into a minimum. This is exemplified in Fig. 3.29 for $a = 2.8$ ($\lambda = 4.6$).

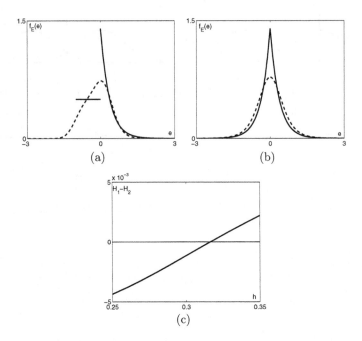

Fig. 3.29 Error PDF models (solid line) of the off-optimal (a) and optimal (b) decision border, respectively $f_1(x; 4.6)$ and $f_2(x; 2.8)$ (see text). The convoluted PDFs (dotted line) were computed with $h = 0.316$, the turn-about value changing the entropy maximum at the optimal border into a minimum. The entropy difference of the convoluted PDFs, as a function of h, is shown in (c) with the zero-crossing at $h = 0.316$.

The maximum-to-minimum effect is general and was observed in large two-class MLPs. It is also observed in multiclass problems for the distinct MLPs outputs, justifying the empirical MEE efficacy in this scenario too [198].

Finally, we show that the change of error entropy behavior as a function of h can also be understood directly from the estimation formulas. Consider, for simplicity, the empirical Rényi's expression, whose minimization as we know is equivalent to the maximization of the empirical information potential

$$\hat{V}_{R_2} = \frac{1}{n^2 h} \sum_{i=1}^{n} \sum_{j=1}^{n} G\left(\frac{e_i - e_j}{h}\right).$$

Let $G_{ij} = G\left(\frac{e_i - e_j}{h}\right)$, $c = \frac{1}{n^2 h}$, and $c_t = \frac{1}{n_t^2 h}$ for $t \in \{-1, 1\}$, where n_t is the number of instances from ω_t. Then, as G is symmetrical, we may write

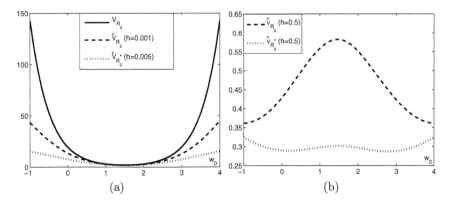

Fig. 3.30 (a) V_{R_2} (solid), $\hat{V}_{R_2}$ (dashed), and $\hat{V}_{R_2^*}$ (dotted) of the data splitter as a function of w_0 for a small value of h; (b) $\hat{V}_{R_2}$ and $\hat{V}_{R_2^*}$ for a higher value of h.

$$\hat{V}_{R_2} = \left(\frac{n_{-1}}{n}\right)^2 c_{-1} \sum_{i\in\omega_{-1}}\sum_{j\in\omega_{-1}} G_{ij} + \left(\frac{n_1}{n}\right)^2 c_1 \sum_{i\in\omega_1}\sum_{j\in\omega_1} G_{ij} + 2c\sum_{i\in\omega_1}\sum_{j\in\omega_{-1}} G_{ij}$$

$$= \hat{q}^2\hat{V}_{R_2|-1} + \hat{p}^2\hat{V}_{R_2|1} + 2c\sum_{i\in\omega_1}\sum_{j\in\omega_{-1}} G_{ij} = \hat{V}_{R_2^*} + 2c\sum_{i\in\omega_1}\sum_{j\in\omega_{-1}} G_{ij}\ .$$

Entropy is therefore decomposed as a weighted sum of positive class-conditional potentials (as in the theoretic derivation of Appendix C), denoted $\hat{V}_{R_2^*}$, plus a term that exclusively relates to cross-errors. Figure 3.30 compares the behavior of V_{R_2}, $\hat{V}_{R_2}$, and $\hat{V}_{R_2^*}$ as a function of the split parameter w_0 for the same problem as in Fig. 3.28. From its inspection we first note the minimum of V_{R_2}, corresponding to the entropy maximum at the optimal solution $w_0^* = 0.5$. Analyzing the behavior of $\hat{V}_{R_2}$, and $\hat{V}_{R_2^*}$ as in Fig. 3.30a it is possible to confirm the convergence of both terms towards V_{R_2} with $h \to 0$. If h is increased above a certain value, $\hat{V}_{R_2}$, and $\hat{V}_{R_2^*}$ will exhibit a maximum at w_0^*, but with an important difference: while the $\hat{V}_{R_2^*}$ maximum is not a global one (for any h), the maximum of $\hat{V}_{R_2}$ is a global one.

Thus, to maximize the empirical information $\hat{V}_{R_2}$, it is important not only to maximize $\hat{q}^2\hat{V}_{R_2|-1} + \hat{p}^2\hat{V}_{R_2|1}$ as for the theoretical counterpart (which can be achieved with different PDF configurations with consequences to the classifier performance; see Property 1 of Sect. 2.3.4) but also to maximize $2c\sum_{i\in\omega_1}\sum_{j\in\omega_{-1}} G_{ij}$ (with no theoretical counterpart), which is achieved if the errors are concentrated at the origin. This cross-error term is due to kernel smoothing.

Note that this analysis also justifies that a larger h is needed to obtain an information maximum for a larger overlap of the class-conditional error PDFs.

Chapter 4
MEE with Discrete Errors

In this chapter we turn our attention to classifiers with a discrete error variable, $E = T - Z$. The need to operate with discrete errors arises when classifiers only produce a discrete output, as for instance the univariate data splitters used by decision trees. For regression-like classifiers, producing Z as a thresholding of a continuous output, $Z = \theta(Y)$, such a need does not arise. The present analysis of MEE with discrete errors, besides complementing our understanding of EE-based classifiers will also serve to lay the foundations of EE-based decision trees later in the chapter.

As before, we focus our attention on the discrete output of simple classifiers, say with $Z \in \{-1, 1\}$ of two-class problems with targets $\{-1, 1\}$. In this framework, the error variable $E = T - Z$ takes value in $\{-2, 0, 2\}$ where $E = -2$ and $E = 2$ correspond to misclassification errors (in class ω_{-1} and ω_1, respectively), while $E = 0$ corresponds to correct classification. The error *probability mass function* (PMF) is therefore written as

$$
P_E(e) = \begin{cases} P_{-1}, & e = -2 \\ 1 - P_{-1} - P_1, & e = 0 \\ P_1, & e = 2 \end{cases} \tag{4.1}
$$

where $P_{-1} = P(E = -2)$ and $P_1 = P(E = 2)$ are the probabilities of error for each class.

Given the nature of the error variable, discrete entropy formulas already introduced in (2.39) and (2.40) must be used. We extensively use the Shannon entropy definition and write:

$$
H_S(E) = -P_{-1} \ln P_{-1} - P_1 \ln P_1 - (1 - P_{-1} - P_1) \ln (1 - P_{-1} - P_1) . \tag{4.2}
$$

In the following we analyze (4.2) as a function of the parameters of the classifier, starting from the special case of a data splitter with univariate input data and then evolving to the more general discrete-output perceptron.

J. Marques de Sá et al.: Minimum Error Entropy Classification, SCI 420, pp. 93–120.
springerlink.com © Springer-Verlag Berlin Heidelberg 2013

4.1   The Data Splitter Setting

The (univariate) discrete data splitter corresponds to a classifier function

$$z = z(x) = \begin{cases} z', & x \leq x' \\ -z', & x > x' \end{cases} , \tag{4.3}$$

where x' is a data split point (or threshold) and $z' \in \{-1, 1\}$ is a class label. The theoretic optimal classification (decision) rule corresponds to a split point x^* and class label z^* such that:

$$(x^*, z^*) = \arg \min P(z(X) \neq t(X)) , \tag{4.4}$$

with

$$\min P_e = \inf \cdot \left\{ \mathbb{1}_{z'=-1} \Big(pF_{X|1}(x') + q(1 - F_{X|-1}(x')) \Big) + \right.$$
$$\left. + \mathbb{1}_{z'=1} \Big(p(1 - F_{X|1}(x')) + qF_{X|-1}(x') \Big) \right\} , \tag{4.5}$$

where $F_{X|t}$ is the distribution function of class ω_t for $t \in \{-1, 1\}$ and p and q are the class priors.. The first term inside braces in equation (4.5) corresponds to the situation where $\min P_e$ is reached when $z' = -1$ is at the left of x'; the second term corresponds to swapping the class labels. A split given by (x^*, z^*) is called a *theoretical Stoller split* [223]. The data-based version, the *empirical Stoller split*, essentially chooses the solution (x', z') such that the empirical error is minimal [223], that is,

$$(x', z') = \arg \min_{(x,z) \in \mathbb{R} \times \{-1,1\}} \frac{1}{n} \sum_{i=1}^{n} \left(\mathbb{1}_{\{X_i \leq x, T_i \neq z\}} + \mathbb{1}_{\{X_i > x, T_i \neq -z\}} \right) . \tag{4.6}$$

The probability of error of the empirical Stoller split converges to the Bayes error for $n \to \infty$ [52].

We assume from now on that ω_{-1} is at the left of ω_1, that is, $z' = -1$, and, our data splitter is given by

$$z = z(x) = \begin{cases} -1, & x \leq x' \\ 1, & x > x' \end{cases} , \tag{4.7}$$

An important result on candidate optimal split points is given by the following theorem [216]:

Theorem 4.1. *For continuous univariate class-conditional PDFs $f_{X|-1}$ and $f_{X|1}$ the Stoller split occurs either at an intersection of $qf_{X|-1}$ with $pf_{X|1}$ or at $+\infty$ or $-\infty$.*

Proof. The probability of error for a given split point x' is given by

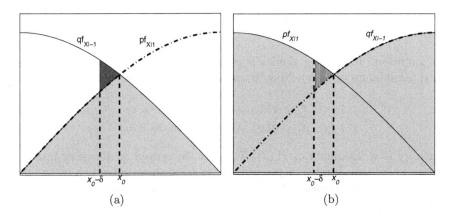

Fig. 4.1 The two intersection cases of $qf_{X|-1}$ and $pf_{X|1}$. The light shadowed areas correspond to $P_e(x_0)$. The dark shadowed area is the amount of error probability added to $P_e(x_0)$ when the split point moves to $x_0 - \delta$. The dashed area is the amount of error probability subtracted from $P_e(x_0)$ when the split point moves to $x_0 - \delta$.

$$P_e(x') = \int_{-\infty}^{x'} pf_{X|1}dx + \int_{x'}^{+\infty} qf_{X|-1}dx. \tag{4.8}$$

If there is no intersection of $qf_{X|-1}$ with $pf_{X|1}$, then $\min P_e = \min(p,q) \le 1/2$ occurs at $+\infty$ or $-\infty$.

For intersecting posterior densities, one has to distinguish two cases. First, assume that for $\delta > 0$

$$pf_{X|1}(x) < qf_{X|-1}(x) \quad x \in [x_0 - \delta, x_0] \tag{4.9}$$
$$pf_{X|1}(x) > qf_{X|-1}(x) \quad x \in [x_0, x_0 + \delta], \tag{4.10}$$

where x_0 is an intersection point of $qf_{X|-1}$ with $pf_{X|1}$ as illustrated in Fig. 4.1a.

The probabilities of error at x_0 and $x_0 - \delta$ are

$$P_e(x_0) = p\left[\int_{-\infty}^{x_0-\delta} f_{X|1}(x)dx + \int_{x_0-\delta}^{x_0} f_{X|1}(x)dx\right] + q\int_{x_0}^{+\infty} f_{X|-1}(x)dx, \tag{4.11}$$

$$P_e(x_0 - \delta) = p\int_{-\infty}^{x_0-\delta} f_{X|1}(x)dx +$$
$$+ q\left[\int_{x_0-\delta}^{x_0} f_{X|-1}(x)dx + \int_{x_0}^{+\infty} f_{X|-1}(x)dx\right]. \tag{4.12}$$

Hence,

$$P_e(x_0) - P_e(x_0 - \delta) = p \int_{x_0-\delta}^{x_0} f_{X|1}(t)dt - q \int_{x_0-\delta}^{x_0} f_{X|-1}(t)dt < 0 , \quad (4.13)$$

by condition (4.9). Using similar arguments, $P_e(x_0) - P_e(x_0 + \delta) < 0$. Thus x_0 is a minimum of $P_e(x)$. Now, suppose that (see Fig. 4.1b)

$$pf_{X|1}(x) > qf_{X|-1}(x) \quad x \in [x_0 - \delta, x_0] \qquad (4.14)$$
$$pf_{X|1}(x) < qf_{X|-1}(x) \quad x \in [x_0, x_0 + \delta] . \qquad (4.15)$$

Then x_0 is a maximum of $P_e(x)$. This can be proven as above or just by noticing that this situation is precisely the same as above but with a relabeling of the classes. For relabeled classes, the respective probability of error $P_e^{(r)}(x)$ is given by

$$P_e^{(r)}(x) = p(1 - F_{X|-1}^{(r)}(x)) + qF_{X|1}^{(r)}(x) =$$
$$= 1 - \left[q(1 - F_{X|-1}(x)) + pF_{X|1}(x)\right] = 1 - P_e(x) . \qquad (4.16)$$

Thus, $P_e^{(r)}(x)$ is just a reflection of $P_e(x)$ around $1/2$, which means that $P_e(x)$ maxima are $P_e^{(r)}(x)$ minima and vice-versa. The optimal split is chosen as the minimum up to a relabeling.   □

Does the solution obtained by minimizing the entropy of the errors produced by a data splitter correspond to the $\min P_e$ solution for the problem? This issue is discussed in the following sections where the error entropy (4.2) is analyzed as a function of the data splitter parameter x', according to the following formulas

$$P_{-1} \equiv P_{-1}(x') = q \left(1 - F_{X|-1}(x')\right),$$
$$P_1 \equiv P_1(x') = p \, F_{X|1}(x'), \qquad (4.17)$$
$$1 - P_{-1} - P_1 = qF_{X|-1}(x') + p \left(1 - F_{X|1}(x')\right) .$$

4.1.1 Motivational Examples

Example 4.1. Consider two uniform distributed classes with PDFs given by

$$f_{X|-1}(x) = \frac{1}{b-a}\mathbb{1}_{[a,b]}(x), \qquad f_{X|1}(x) = \frac{1}{d-c}\mathbb{1}_{[c,d]}(x) , \qquad (4.18)$$

such that $a < c$ (ω_{-1} is at the left of ω_1). If the classes are separable, that is $a < b < c < d$, the $\min P_e$ solution is obtained if x^* is set anywhere in $]b, c[$. The minimum entropy value $H_S(E) = 0$ also occurs in that interval because $P(E = 0) = 1$. So, minimizing Shannon's entropy of the error (SEE) leads in this case to the $\min P_e$ solution.

For overlapped class distributions, such that $a < c \le b < d$, one computes

$$
P_e = P_{-1} + P_1 = \begin{cases}
q, & x' < a \\
q\,\frac{b-x'}{b-a}, & x' \in [a, c[\\
q\,\frac{b-x'}{b-a} + p\,\frac{x'-c}{d-c}, & x' \in [c, b[\\
p\,\frac{x'-c}{d-c}, & x' \in [b, d[\\
p, & x' \ge d
\end{cases}
\tag{4.19}
$$

and

$$
H_S(E) = \begin{cases}
q \ln q + p \ln p, & x' < a \\
q\frac{b-x'}{b-a} + \left(q\frac{x'-a}{b-a} + p\right)\ln\left(q\frac{x'-a}{b-a} + p\right), & x' \in [a, c[\\
q\frac{b-x'}{b-a}\ln\left(q\frac{b-x'}{b-a}\right) + p\frac{x'-c}{d-c}\ln\left(p\frac{x'-c}{d-c}\right) + \\
\quad + \left(q\frac{x'-a}{b-a} + p\frac{d-x'}{d-c}\right)\ln\left(q\frac{x'-a}{b-a} + p\frac{d-x'}{d-c}\right), & x' \in [c, b[\\
p\frac{x'-c}{d-c}\ln\left(p\frac{x'-c}{d-c}\right) + \left(q + p\frac{d-x'}{d-c}\right)\ln\left(q + p\frac{d-x'}{d-c}\right), & x' \in [b, d[\\
p \ln p + q \ln q, & x' \ge d
\end{cases}
\tag{4.20}
$$

where we have used formulas (4.2) and (4.17) and the usual convention $0 \ln 0 = 0$.

Note that both P_e and H_S are functions of x' although we omit this dependency for notational simplicity.

Figure 4.2 shows P_e and H_S as functions of x' for two different settings. In Fig. 4.2a min P_e is attained for x^* anywhere in $[0.7, 1]$ (where P_e is constant), but error entropy attains its minimum at the extremes of the overlapped region, $x^* = 0.7$ or $x^* = 1$. The reason lies in the decrease of uncertainty (and consequently of entropy) experimented by the error variable taking only two values (in $\{-2, 0\}$ or $\{0, 2\}$) for those choices of x^*: entropy prefers to correctly classify one class at the expense of the other. This behavior is general for any class setting with $b - a = d - c$. Figure 4.2b illustrates a more general setting, but again the global minimum of $H_S(E)$ matches the min P_e solution. This is observed for any class setting, which means that Shannon error entropy splits are always optimal for uniform distributed classes (for a complete proof see [216]).   □

Example 4.2. Consider two Gaussian classes with mean μ_t and standard deviations σ_t, for $t \in \{-1, 1\}$. Then

$$
F_{X|t}(x') = \int_{-\infty}^{x'} \frac{1}{\sqrt{2\pi}\sigma_t} e^{-\frac{(x-\mu_t)^2}{2\sigma_t^2}} = \Phi\left(\frac{x' - \mu_t}{\sigma_t}\right)
\tag{4.21}
$$

where $\Phi(\cdot)$ is the standardized Gaussian CDF. Figure 4.3 shows $H_S(E)$ and P_e as a function of the split point x', revealing that entropy has a quite different behavior depending on the class settings. First, Fig. 4.3a, 4.3b and

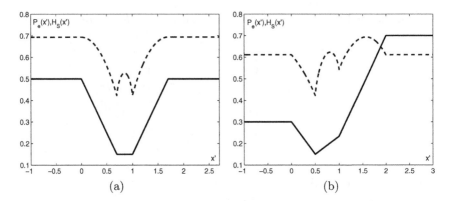

Fig. 4.2 H_S (dashed line) and P_e (solid line) plotted as functions of x' in a two-class Uniform problem: a) $[a,b] = [0,1]$, $[c,d] = [0.7, 1.7]$ and $p = q = 0.5$; b) $[a,b] = [0,1]$, $[c,d] = [0.5, 2]$ and $p = 0.7$.

4.3c (and similar ones) suggest that the correspondence between the MEE solution and the min P_e solution depend on some sort of $q f_{X|-1}$ vs $p f_{X|1}$ balance between the classes. Second, the min P_e solution may correspond to an entropy maximum depending on the distance between the classes as illustrated in Figs. 4.3a and 4.3d.   □

4.1.2 SEE Critical Points

Our first goal is to compute the several risks described in Chap. 2, using formulas (4.17) and analyze their critical points by relating them to the min P_e solution. We then focus our analysis in SEE seeking a theoretical explanation for the behavior encountered in Example 4.2.

Let $p = q = 1/2$ and recall from Theorem 4.1 that at the min P_e solution $f_{X|-1}(x^*) = f_{X|1}(x^*)$. Then:

1. MSE risk

$$R_{MSE}(x') = \sum_e e^2 P_E(e) = (-2)^2 P_{-1} + (0)^2 (1 - P_{-1} - P_1) + (2)^2 P_1$$

$$= 4(P_{-1} + P_1) = 4P_e. \tag{4.22}$$

We see that $R_{MSE}(x')$ depends, in this case, only on the misclassified instances. In order to determine x^*, first note that by the Fundamental Theorem of Calculus $dF_{X|t}/dx' = f_{X|t}(x')$. Therefore:

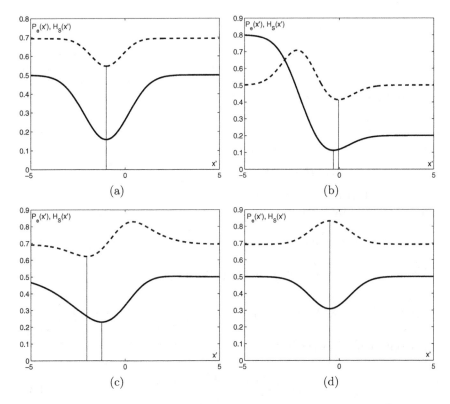

Fig. 4.3 H_S (dashed line) and P_e (solid line) plotted as functions of x' in a two-class Gaussian problem. Corresponding minima (maximum for H_S in d)) are marked with vertical lines. Taking a) as reference where $\mu_{-1} = -2$, $\mu_1 = 0$, $\sigma_{-1} = \sigma_1 = 1$ and $p = q = 0.5$, one has: b) $p = 0.2$; c) $\sigma_{-1} = 2$; d) $\mu_{-1} = -1$.

$$\frac{d}{dx'} R_{MSE}(x') = 0 \;\Rightarrow\; f_{X|-1}(x') = f_{X|1}(x') . \qquad (4.23)$$

Thus, the MMSE procedure produces an optimum x' corresponding to the intersection of $f_{X|-1}$ with $f_{X|1}$, which happens to be the min P_e solution.

2. Quadratic Rényi EE risk

$$R_{R_2 EE}(x') \equiv H_{R_2}(x') = -\ln\left(\sum_e P_E^2(e)\right) . \qquad (4.24)$$

The information potential is then

$$V_{R_2}(x') = \exp(-H_{R_2}(x')) = P_{-1}^2 + (1 - P_{-1} - P_1)^2 + P_1^2 \qquad (4.25)$$

whose maximizing x' will minimize $H_{R_2}(x')$. We have:

$$\frac{d}{dx'}V_{R_2}(x') = -P_{-1}f_{X|-1}(x') +$$
$$+ (1 - P_{-1} - P_1)(f_{X|-1}(x') - f_{X|1}(x')) + P_1 f_{X|1}(x') .$$
$$(4.26)$$

The optimum x' satisfies

$$\frac{f_{X|-1}(x')}{f_{X|1}(x')} = \frac{1 - 2P_1 - P_{-1}}{1 - 2P_{-1} - P_1} \qquad (4.27)$$

implying at x^* that $P_{-1} = P_1$.

3. Shannon EE risk
   R_{SEE} is given by formula (4.2)

$$R_{SEE}(x') = -P_{-1}\ln P_{-1} - P_1 \ln P_1 - (1 - P_{-1} - P_1)\ln(1 - P_{-1} - P_1) .$$
$$(4.28)$$

Therefore, after some simple manipulations:

$$\frac{d}{dx'}R_{SEE}(x') = \frac{1}{2}f_{X|-1}(x')\ln\frac{P_{-1}}{1 - P_{-1} - P_1} - \frac{1}{2}f_{X|1}(x')\ln\frac{P_1}{1 - P_{-1} - P_1} .$$
$$(4.29)$$

The optimum x' satisfies

$$\frac{f_{X|1}(x')}{f_{X|-1}(x')} = \frac{\ln\frac{P_{-1}}{1-P_{-1}-P_1}}{\ln\frac{P_1}{1-P_{-1}-P_1}} \qquad (4.30)$$

implying at x^* that $P_{-1} = P_1$.

This result for SEE can in fact be easily proved for any $p = 1 - q$ setting, constituting the following theorem whose proof can be found in [216]:

Theorem 4.2. *In the univariate two-class problem with continuous class-conditional PDFs, the* $\min P_e$ *solution* x^* *is a critical point of the Shannon error entropy if the error probabilities of each class at* x^* *are equal.* □

This result justifies part of what was observed in Example 4.2. If the classes are not balanced, in the sense of equal error probability, then x^* is not a *critical point* of SEE. However, Theorem 4.2 says nothing about the nature (maximum or minimum) of such critical points. Indeed, as Fig. 4.3d has shown, the obtained solution is not guaranteed to be an entropy minimum.

   Before performing a more analytical treatment, Fig. 4.4 gives us an intuitive explanation for the fact. If the classes are sufficiently distant $P_E(e)$ exhibits a peaky PMF at x^* (almost all probability is concentrated in $P(E = 0)$), close to a Dirac-δ. Therefore, SEE is lower at x^* (see the top row of Fig. 4.4). On the other hand, when the classes get closer and the amount of class overlap increases the PMF peakedness is gradually lost. At

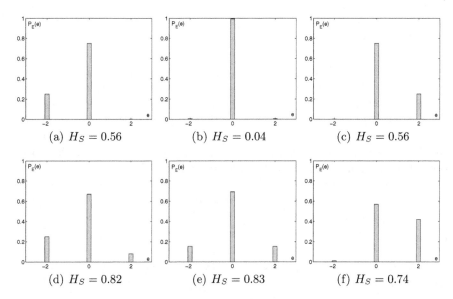

(a) $H_S = 0.56$ (b) $H_S = 0.04$ (c) $H_S = 0.56$

(d) $H_S = 0.82$ (e) $H_S = 0.83$ (f) $H_S = 0.74$

Fig. 4.4 Probability mass functions for distant (top) and close (bottom) classes in the Stoller split setting. Figures from left to right correspond to the split position at the left, at the location and at the right of the optimal split, respectively.

x^*, $P_E(e)$ has more entropy than at any other split point (see the bottom row of Fig. 4.4).

This behavior, observed for Gaussian classes, is in fact quite general, as the following discussion demonstrates. The sign of $\frac{d^2 H_S}{dx'^2}(x^*)$ (and thus, the nature of the critical point x^*) can be analyzed as a function of the distance between the classes, defined as the distance d between their centers (medians). For that, and w.l.o.g., let us consider that class ω_1 is centered at 0 and the center of class ω_{-1} moves along the non-positive side of the real line. Then, $x^* = -d/2$. Now,

$$\frac{d^2 H_S}{dx'^2}(x') = q\frac{df_{X|-1}}{dx'}(x') \ln\left(\frac{P_{-1}}{1 - P_{-1} - P_1}\right) -$$
$$- p\frac{df_{X|1}}{dx'}(x') \ln\left(\frac{P_1}{1 - P_{-1} - P_1}\right) -$$
$$- \frac{(qf_{X|-1}(x') - pf_{X|1}(x'))^2}{1 - P_{-1} - P_1} - \frac{q^2 f_{X|-1}^2(x')}{P_{-1}} - \frac{p^2 f_{X|1}^2(x')}{P_1}.$$

$$(4.31)$$

From Theorems 4.1 and 4.2 one has $pf_{X|1}(x^*) = qf_{X|-1}(x^*)$ and $P_{-1}(x^*) = P_1(x^*)$ therefore

$$\frac{d^2 H_S}{dx'^2}(x^*) = \ln\left(\frac{P_1}{1 - 2P_1}\right)\left[q\frac{df_{X|-1}}{dx'}(x^*) - p\frac{df_{X|1}}{dx'}(x^*)\right] - \frac{2p^2 f_{X|1}^2(x^*)}{P_1} .$$
$$(4.32)$$

This expression is quite intractable as it involves general class-conditional PDFs. Important simplifications can be made by considering the case of mutually symmetric class distributions [216], that is, where $qf_{X|-1}(x)$ has the same shape of $pf_{X|1}(-x)$. We readily see that $P_{-1}(x^*) = P_1(x^*)$ for this type of two-class problems. In particular, the Gaussian two-class problem with equal standard deviations and priors is an example of such type of distributions. For mutually symmetric class distributions one has

$$q\frac{df_{X|-1}}{dx'}(x^*) = -p\frac{df_{X|1}}{dx'}(x^*) ;$$
$$(4.33)$$

hence,

$$\frac{d^2 H_S}{dx'^2}(x^*) = -2\left(p\frac{df_{X|1}}{dx'}(x^*)\ln\frac{P_1}{1 - 2P_1} + \frac{p^2 f_{X|1}^2(x^*)}{P_1}\right) .$$
$$(4.34)$$

Let

$$Q(x^*) = \frac{df}{dx'}\ln\frac{P}{1 - 2P} + \frac{f^2}{P} ,$$
$$(4.35)$$

where for notation simplicity we took $f \equiv pf_{X|1}(x^*)$ and $P \equiv P_1(x^*)$. The function $Q(x^*)$ plays a key role in the analysis of the error entropy critical points as carefully discussed in [216]. In fact, for increasingly distant classes (increasing d with $x^* \to -\infty$) $Q(x^*)$ can be shown to be negative and thus error entropy has a minimum at x^*. On the other hand, if the classes get closer (decreasing d with $x^* \to 0$) $Q(x^*)$ may change sign and thus x^* turns out to be a maximizer of error entropy. Such is the case when the mutually symmetric class distributions are by themselves symmetric like in Gaussian-classes problems.

Summarizing, for a large class of problems optimization of SEE performs optimaly whether in the sense of minimization (MEE) for sufficiently separated classes or maximization for close classes. Of course, it would also be relevant to know when to choose between optimization strategy, that is, to know what is the minimum-to-maximum turn-about distance between the classes. Unfortunately, the answer has to be searched case by case for each pair of mutually symmetric distributions, taking into account that for a scaled version of X, $Y = \Delta X$, the ratio x^*/Δ between the $Q(x^*) = 0$ solution and the scale Δ, is a constant. In fact one has

$$F_Y(y) = F_X(y/\Delta),$$

$$f_Y(y^*) = \frac{1}{\Delta} f_X(y^*/\Delta),$$ (4.36)

$$\frac{df_Y}{dy}(y^*) = \left(\frac{1}{\Delta}\right)^2 \frac{df_X}{dx}(y^*/\Delta).$$

Therefore,

$$Q(y^*) = \frac{df_Y}{dy'}(y^*) \ln \frac{F_Y(y^*)}{1 - 2F_Y(y^*)} + \frac{f_Y^2(y^*)}{F_Y(y^*)} = 0 \Leftrightarrow$$

$$\Leftrightarrow \left(\frac{1}{\Delta}\right)^2 \left(\frac{df_X}{dx'}(y^*/\Delta) \ln \frac{F_X(y^*/\Delta)}{1 - 2F_X(y^*/\Delta)} + \frac{f_X^2(y^*/\Delta)}{F_X(y^*/\Delta)}\right) = 0 \Leftrightarrow$$

$$\Leftrightarrow \left(\frac{1}{\Delta}\right)^2 Q(x^*) = 0.$$ (4.37)

Hence, the solution is $y^* = \Delta \cdot x^*$.

The turn-about value t_{value} for symmetric triangular distributions is $t_{value} = 0.3473\Delta$ (see [216] for details). In the following examples we analyze the t_{value} for the case of mutually symmetric Gaussian and lognormal class distributions.

Example 4.3. Consider two Gaussian distributed classes with scale $\Delta = \sigma_{\pm 1} = \sigma = 1$ and distance d between their centers (means). Setting $p = q = 1/2$, $Q(x^*)$ can easily be re-written as a function of d as follows

$$Q(d) = \frac{d}{4\sqrt{2\pi}} \exp(-d^2/8) \ln \left(\frac{1 - \Phi(d/2)}{2\Phi(d/2)}\right) + \frac{\exp\left(-d^2/4\right)}{4\pi(1 - \Phi(d/2))},$$ (4.38)

and

$$Q(d) = 0 \Leftrightarrow \frac{d}{2} \ln \left(\frac{1 - \Phi(d/2)}{2\Phi(d/2)}\right)(1 - \Phi(d/2)) + \frac{\exp\left(-d^2/8\right)}{\sqrt{2\pi}} = 0 \Leftrightarrow$$

$$\Leftrightarrow d = 1.4052.$$ (4.39)

Thus, from the preceding discussion one may conclude that the turn-about value t_{value} for a Gaussian two-class problem with scale σ is $t_{value} = 1.4052\,\sigma$, that is, one should maximize SEE when $d < t_{value}$.   □

Example 4.4. Consider now the lognormal distribution with PDF

$$f_X(x) = \frac{1}{x\sigma\sqrt{2\pi}} \exp\left(-\frac{(\ln x - \mu)^2}{2\sigma^2}\right), \quad x > 0.$$ (4.40)

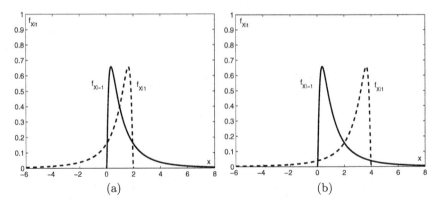

Fig. 4.5 The lognormal two-class problem for different values of d: a) $d = 0$; b) $d = 2$.

Our two-class problem is defined by the following input PDFs

$$f_{X|-1}(x) = \frac{1}{x\sqrt{2\pi}} \exp\left(-\frac{\ln^2(x)}{2}\right) \ x > 0, \text{(that is, } \mu = 0, \sigma = 1)\text{ ,} \quad (4.41)$$

$$f_{X|1}(x) = f_{X|-1}(-x + 2 + d) \quad (4.42)$$

where d is a non-negative real number, measuring the distance between the class centers (medians). Figure 4.5 shows this two-class problem for two different values of d. We focus on the inner intersection as solution to our problem. Figure 4.6 shows $Q(d)$ (we omitt the expression as it is rather intricate) with its $Q(d) = 0$ solution of $d = 0.124$. Note, however, that for such small values of d, the Stoller split is no longer at the inner intersection, but at one of the outer intersections of the class conditional PDFs with a relabel of the classes (the Stoller split is at the inner intersection with the usual left ω_{-1} and right ω_1 order of the classes for $d \gtrsim 0.5$). That is, as long as the Stoller split is at the inner intersection, then it corresponds to a SEE minimum.   □

4.1.3 Empirical SEE Splits

Theoretical SEE splits are only possible to obtain if the class distributions are known. We now present two methods to empirically deal with SEE when the class distributions are unknown. The first method uses the KDE estimates of the input distributions as a means to compute P_{-1} and P_1. The second method estimates the same quantities directly from the misclassified samples.

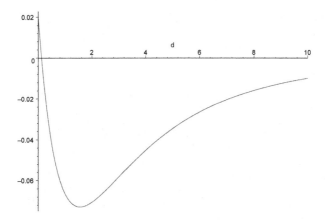

Fig. 4.6 The function $Q(d)$ for the lognormal two-class problem.

4.1.3.1 Kernel-Based Estimates

Recall that

$$P_{-1} = \int_{x'}^{-\infty} q f_{X|-1}(x)dx = q \left[1 - \int_{-\infty}^{x'} f_{X|-1}(x)dx \right], \qquad (4.43)$$

$$P_1 = p \int_{-\infty}^{x'} f_{X|1}(x)dx . \qquad (4.44)$$

Using the KDE (3.2) to estimate $f_{X|t}$ one gets

$$\int_{-\infty}^{x'} f_{X|t}(x)dx \approx \frac{1}{n} \sum_{x_i \in \omega_t} \Phi \left(\frac{x' - x_i}{h} \right) . \qquad (4.45)$$

An important difference with respect to the empirical version for continuous errors is to be noticed. Here KDE is used to estimate the input PDFs in contrast with the estimation of the error PDF performed in the previous chapter.

An appropriate optimization algorithm can now be used to obtain SEE solutions. We illustrate this procedure with simulated data, namely two Gaussian distributed classes satisfying the conditions of Theorem 4.2: $p = q = 1/2$ and $\sigma_{-1} = \sigma_1$. As discussed in the previous section, one may set $\sigma_t = 1$ and vary the distance d between the classes. Given that $t_{value} \approx 1.405$, we performed experiments with generated data using $d = 3$, $d = 1.5$ and $d = 1$, the first two values corresponding to minimization problems and the third one to a maximization problem. The experiments use different values of the training set and test set sizes shown in Tables 4.1 and 4.2: n_d, n_t instances, equally divided between the two classes. Adequate values of h can be determined taking formula (E.19) into account: for 100 instances per class, $h_{IMSE} = 0.42$.

Table 4.1 Percentage of test error and standard deviations (in parentheses) obtained with SEE for the simulated Gaussian data. Minimization was used for distances $d = 3$ and $d = 1.5$, and maximization for $d = 1$.

				n_d	
d	Bayes error	n_t	200	2000	20000
		10^2	6.79(2.41)	6.75(2.51)	6.75(2.51)
3	6.68%	10^4	6.82(0.83)	6.70(0.81)	6.66(0.81)
		10^6	6.81(0.20)	6.69(0.08)	6.68(0.08)
		10^2	25.23(4.65)	24.67(4.58)	22.61(4.21)
1.5	22.66 %	10^4	25.32(2.49)	24.72(2.15)	22.80(0.46)
		10^6	25.46(2.54)	24.83(2.21)	22.82(0.24)
		10^2	30.63(4.64)	30.90(4.48)	30.70(4.82)
1	30.85%	10^4	30.93(0.47)	30.87(0.47)	30.84(0.46)
		10^6	30.93(0.17)	30.86(0.14)	30.85(0.14)

Table 4.2 Percentage of test error and standard deviation (in parentheses) obtained with maximization of SEE with increased h ($h = 2.27$) for $d = 1.5$.

				n_d	
d	Bayes error	n_t	200	2000	20000
		10^2	22.95(3.93)	22.78(4.01)	22.47(4.14)
1.5	22.66%	10^4	22.73(0.41)	22.65(0.43)	22.66(0.41)
		10^6	22.75(0.17)	22.67(0.14)	22.67(0.13)

Table 4.1 shows the mean values and standard deviations over 1000 repetitions for the test error of each experiment, using $h = 1.7$, 0.1 and 0.8 for $d = 1$, 1.5 and 3, respectively. When the amount of available data is huge, SEE achieves Bayes discrimination as expected. For small datasets the picture is quite different. SEE still finds a good solution for $d = 3$ and $d = 1$, but performs poorly for $d = 1.5$, which is near the t_{value} for Gaussian classes, that is, in the limbo between a choice to minimize or maximize. The reason lies in the highly non smooth estimate of the input distributions by the KDE method, with $h < h_{IMSE}$. This can be solved by using fat estimation of the input PDFs as we did for $d = 1$ and $d = 3$. As shown in Fig. 4.7, when h is too small one gets a non-smooth entropy function while for large h the oversmoothed input PDF estimates provide a smooth entropy curve preserving the maximum. The results of Table 4.2 were obtained by using fat estimation ($h = 2.27$) to the case $d = 1.5$. Now, SEE performs similarly as in the cases $d = 1$ and $d = 3$.

The explanation for this behavior relies on the increased variance of the estimated PDFs, which for a Gaussian kernel is given by $\hat{\sigma} = s^2 + h^2$ (see

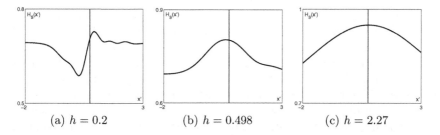

(a) $h = 0.2$ (b) $h = 0.498$ (c) $h = 2.27$

Fig. 4.7 Error entropy for different values of h in the Gaussian distribution example of Sect. 4.1.3.1 with $d = 1.5$. The location of the optimal solution is marked with a vertical line.

formula (E.5)). In practice, the increased h has the effect of approximating the classes (recall the discussion at the end of the previous section) turning the minimization problem into a maximization one. Of course, one should not increase h indefinitely, because the optimization algorithm would make gross mistakes when dealing with an almost flat entropy curve. A discussion on this issue as well as additional experiments with real world datasets can be found in [212, 216].

4.1.3.2 Resubstitution Estimates

The kernel-based estimation method is not an attractive method to be used in practice due to its computational burden. It is far simpler to compute the empirical estimate of (4.2) using the usual resubstitution estimation of the probabilities. For notational simplicity sake we will use in this and the following sections 0-1 class labeling ($T = \{0, 1\}$), and write the empirical estimate of (4.2) as

$$SEE \equiv SEE(\hat{P}_{01}, \hat{P}_{10}) = -\hat{P}_{01} \ln \hat{P}_{01} - \hat{P}_{10} \ln \hat{P}_{10} -$$
$$- (1 - \hat{P}_{01} - \hat{P}_{10}) \ln(1 - \hat{P}_{01} - \hat{P}_{10}) \qquad (4.46)$$

with

$$\hat{P}_{10} \equiv \hat{P}(E = 1) = \hat{P}(T = 1, Y = 0), \text{ the error probability of class } \omega_k,$$
$$\hat{P}_{01} \equiv \hat{P}(E = -1) = \hat{P}(T = 0, Y = 1), \text{ the error probability of class } \overline{\omega}_k ,$$

where ω_k, $k = 1, \ldots, c$, is our class of interest in some two-class discrimination problem, which we want to discriminate from its complement $\overline{\omega}_k$.

The resubstitution estimate of SEE uses the error rate estimates $\hat{P}_{10} = n_{10}/n$, $\hat{P}_{01} = n_{01}/n$ with $n_{tt'}$ denoting the number of class t cases classified as t'.

We have seen that the theoretical min P_e point in two-class univariate problems coincides or is in a close neighborhood of the MEE point when the class conditional distributions are not too overlapped, and corresponds to max P_e otherwise, with the min-max turn-about value depending on the distributions. The empirical MEE point also displays the same behavior, as illustrated in Fig. 4.8 where SEE is shown for two different feature-class combinations of the well-known Iris dataset [13]. In Fig. 4.8a the distribution overlap is small and the MEE split point occurs close to the min P_e point. In Fig. 4.8b, with large distribution overlap, the MEE split point occurs at an end of the variable spanned interval, whereas the min P_e point occurs in the vicinity of max SEE.

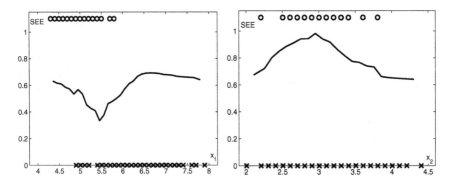

Fig. 4.8 SEE curves for two splits of the Iris dataset (splitting the balls from the crosses): a) class 1 (Iris setosa), feature x_1 (sepal length): MEE at $x_1 = 5.45$; b) class 3 (Iris Virginica), feature x_2 (sepal width): MEE at $x_2 = 4.3$.

We will see later how to capitalize on the apparently annoying fact that the MEE split point occurs at an end of the variable spanned interval for overlapped distributions. As a matter of fact, we will use this interval-end criterion as a synonym of "overlapped".

Experimental studies consisting of applying the empirical MEE procedure to artificially generated datasets, with known mutually symmetric distributions, are expected to confirm the theoretical findings of the preceding Sect. 4.1.2 and provide further evidence regarding the interval-end criterion we mentioned. One such study was carried out for classes with Gaussian distributions of the data instances in [152]. We present here a few more results obtained following the same procedure as in the cited work, which consisted of measuring the error rate and the interval-end hit rate of the empirical MEE (for SEE) split point for equal-variance Gaussian distributed data. Concretely, setting $\sigma_t = 1$ and $\mu_0 = 0$, we varied μ_1 in a grid of points and generated n normally distributed instances for both classes with those parameters.

For a grid of μ_1 values in $[0, 4]$ and a variable number of instances per class, n, 400 experiments were performed by generating the class distributions with the above settings and determining the empirical MEE point; this allowed determining for each μ_1 value the percentage of MEE points falling at an interval end, and the average and standard deviation of the error rate. For $n = 120$ instances per class, Fig. 4.9a shows the interval-end hit rate; Fig. 4.9b shows the average error rate ($\pm$ standard deviation) with the superimposed curve of min P_e (the Bayes error rate) computed with formula (3.34).

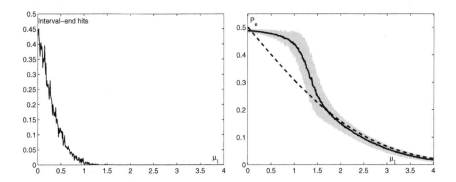

Fig. 4.9 Empirical MEE for a two-class setting with Gaussian distributions, $g_0(x; 0, 1)$ and $g_1(x; \mu_1, 1)$: a) interval-end hit rate; b) Average error rate (solid line) with $\pm$standard deviation (shadowed area) with the theoretical min P_e (dotted line).

From this and other experiments with smaller values of n, the following conclusions can be drawn:

1. The experimental findings confirm the theoretical turn-about value for symmetrical Gaussian distributions (Sect. 4.1.2), corresponding here to $\mu_1^+ = 1.405$. Below μ_1^+ the interval-end hit rate increases steadily, as shown in Fig. 4.9a, reaching values near 50% or even above (for small values of n). The average error rate also exhibits the turn-about effect: the estimated derivative (using a moving-average filtered version of the average error rate, with filter lengths of 3 to 9), exhibits an inflection point at 1.41, close to μ_1^+. For smaller values of n the inflected aspect of the average error rate is also present but in a less pronounced way.
2. Above μ_1^+ the average error rate is close to the Bayes error with fast decreasing standard deviation. This behavior is also observed for small values of n.

The same experiments carried out on mutually symmetric triangular distributions with unit interval ($b - a = 1$; see 2.47) and lognormal distributions with $\sigma = 0.5$ lead to the same conclusions. Figure 4.10 shows the results for these distributions obtained in 400 experiments with $n = 50$. The empirical

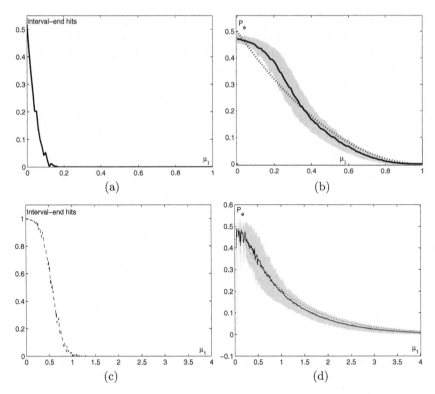

Fig. 4.10 Interval-end hit rates and average error rates, represented as in Fig. 4.9, for unit interval triangular distributions — top plots (a) and (b) and lognormal distributions with $\sigma = 0.5$ — bottom plots (c) and (d).

turn-about point, calculated as before, for triangular and lognorm distributions is in close agreement with the theoretical values of 0.347 and 0.362, respectively. Note that in Example 4.4 the turn-about value for the lognorm distribution with $sigma = 1$ was 0.124 ; for $sigma = 0.5$, one has 0.362. The mutually symmetric lognormal case is interesting because it exemplifies for moderately small μ_1 the severe case of distribution overlapping with an interval-end hit rate reaching 100%.

4.1.4   MEE Splits and Classic Splits

Univariate splits are the basic ingredient in the construction of binary decision trees. Tree classifiers use consecutive input domain partitions by applying decision rules at tree nodes until reaching a so-called tree leaf corresponding to a tree output. Denoting any tree node by u, a univariate split at any tree node u_j, represents a binary test z_j as $\{x_{ij} \leq \Delta,\ z_j(x_i) = \omega_k;\ \overline{\omega}_k\ \text{otherwise}\}$ for *numerical* inputs (i.e., real-valued inputs) or as $\{x_{ij} \in B,\ z_j(x_i) = \omega_k;\ \overline{\omega}_k,\ \text{otherwise}\}$ for *categorical* inputs; Δ and B are, respectively, a numerical threshold and a set of categories. The application of the binary test at a tree node u sends the inputted data either to a left node u_l — if $x_{ij} \leq \Delta$ or $x_{ij} \in B$ —; or to a right node u_r — if $x_{ij} > \Delta$ or $x_{ij} \notin B$. The left and right nodes, u_l and u_r, are called children nodes of u (the parent node). Thus, the application of successive binary tests sends a data instance traveling down the tree until reaching some tree leaf, where a final class label is assigned.

Details on the construction of binary trees can be found in [33, 188]. In the following we describe three classic univariate splits, often used in the construction of binary decision trees, and compare them with SEE splits [152].

4.1.4.1   Classic Splitting Criteria

The many univariate splits (also named splitting criteria or rules) proposed until today fall into two main categories: impurity-based criteria and binary criteria [188, 147]. All of them are defined in terms of the class-conditional PMFs, $P(\omega_i|u)$ at a node u. In contrast, MEE is *not* defined in terms of the $P(\omega_i|u)$; MEE is *post-rule* defined, in terms of the $P(\omega_i|r)$, where r denotes a splitting rule.

An impurity function of a node u, $\phi(u) \equiv \phi(P(\omega_1|u), \ldots, P(\omega_c|u))$, is a symmetric function measuring the degree of "disorder" of its PMF in such a way that it achieves its maximum at the uniform distribution, and the minimum at a Dirac-*delta* distribution. Classic impurity functions are the Gini diversity index, $GDI(u) = \sum_{j=1}^{c} \sum_{k=1, k \neq j}^{c} P(\omega_j|u)P(\omega_k|u)$ and, of course, the (discrete) Shannon entropy (also called "information" in decision tree jargon) $H(u) = -\sum_{k=1}^{c} P(\omega_k|u) \ln P(\omega_k|u)$.

Consider a node u, candidate to splitting into right and left nodes, u_r and u_l. The average impurity is $\overline{\phi}_z(u) = P(u_r|u, z)\phi(u_r) + P(u_l|u, z)\phi(u_l)$. One is then interested in maximizing $\Delta_z\phi(u) = \phi(u) - \overline{\phi}_z(u)$ at each node. When using $GDI(u)$ and $H(u)$ we obtain the Gini index (GI) and information gain (IG) split criteria, respectively. Since $GDI(u)$ is the first order approximation of $H(u)$, both GI and IG behave similarly, allowing us to solely focus on IG in the sequel [35, 178].

The probability of error at u, $PE(u) = 1 - \max_j P(\omega_j|u)$, is also an impurity function, since it enjoys all the above properties. However, as shown in [33], for tree design $PE(u)$ has two serious drawbacks: it may lead to $\omega(u) = \omega(u_r) = \omega(u_l)$ for all candidate splits and in such cases no best split can be found; it may lead to reject solutions with purer children nodes in detriment of others with highly impure nodes, the reason for this shortcoming being the non-strict concave behavior of $PE(\cdot)$ (*GDI* and H are strictly concave functions).

SEE is *not* a node impurity function. Whereas a node impurity function depends solely on the $P(\omega_i|u)$, SEE depends on the "error" of a specific split involving parent and children nodes. Moreover, whereas node impurity functions have a c-dimensional support for any arbitrary c, SEE always has a 3-point support.

Binary splitting criteria are based on partitions of feature domains into two sub-domains. The classic binary criterion is the "towing" splitting rule. Corresponds to maximizing $TWO(u) = P(u_r|u)P(u_l|u)[\sum_{\omega_i \in \Omega}|P(\omega_i|u_r) - P(\omega_i|u_l)|]^2$, measuring how far apart are the numbers of cases of each class assigned to each of the two subdomains, u_r and u_l.

4.1.4.2 Comparing IG, TWO, PE, and SEE

The comparative study of the four splitting criteria requires a common framework expressing them all uniformly. We start by expressing all criteria in terms of a minimization problem (one could of course express them all as a maximization problem). For the impurity criterion $IG(u)$ (and $GI(u)$), since $\phi(u)$ doesn't depend on the rule, we minimize $\overline{\phi}_z(u)$ instead of maximizing $\Delta_z\phi(u)$; for the towing criterion we minimize $\max(TWO(u)) - TWO(u)$. For simplicity reasons we keep the names $IG(u)$ and $TWO(u)$ for the min-expressed rules.

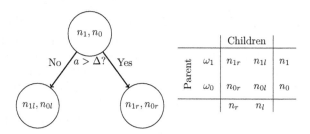

Fig. 4.11 Two-class framework.

Next, we consider two classes, ω_1 and $\omega_0 \equiv \overline{\omega}_1$, and the splitting of a node with n_1 instances from class 1 and n_0 from class 0 as shown in Fig. 4.11. At the parent node, with $n = n_1 + n_0$ instances, we have: $P(\omega_1) = n_1/n = p$; $P(\omega_0) = n_0/n = 1 - p = q$. Moreover, for any prior p and whatever criterion being used, we consider that a certain percentage α of ω_1 instances are assigned to the right node and a certain percentage β of ω_0 instances are assigned to the left node: $n_{1r} = \alpha n_1$; $n_{0l} = \beta n_0$. In other words, assuming that the labels assigned to the right and left nodes are respectively ω_1 and ω_0, α and β represent the percentages of correctly classified instances. With these provisions we achieve what we were looking for: all splitting rules are expressed in terms of the minimization of a function depending on the same parameters, p, α and β (the least possible number of parameters), as follows:

- **Information-Gain criterion**

  Denoting $n_r = n_{1r} + n_{0r}$, $n_l = n_{1l} + n_{0l}$, we have:

  Entropy of the right node: $H_r = -\dfrac{n_{1r}}{n_r} \ln \dfrac{n_{1r}}{n_r} - \dfrac{n_{0r}}{n_r} \ln \dfrac{n_{0r}}{n_r}$.

  Entropy of the left node: $H_l = -\dfrac{n_{1l}}{n_l} \ln \dfrac{n_{1l}}{n_l} - \dfrac{n_{0l}}{n_l} \ln \dfrac{n_{0l}}{n_l}$.

  Average entropy to be minimized: $IG = \dfrac{n_r}{n} H_r + \dfrac{n_l}{n} H_l$.

  Denoting $\overline{\alpha} = 1 - \alpha$ and $\overline{\beta} = 1 - \beta$, one derives after simple manipulations:

  $$IG(p, \alpha, \beta) = -\alpha p \ln \frac{\alpha p}{\alpha p + \overline{\beta} q} - \overline{\beta} q \ln \frac{\overline{\beta} q}{\alpha p + \overline{\beta} q} -$$
  $$\overline{\alpha} p \ln \frac{\overline{\alpha} p}{\overline{\alpha} p + \beta q} - \beta q \ln \frac{\beta q}{\overline{\alpha} p + \beta q} \ . \quad (4.47)$$

  Formula 4.47 clearly shows that $IG(p, \alpha, \beta)$ is *independent* of the predicted class labeling; swapping α by β together with p by q in formula 4.47 leads to the same result.

- **Twoing criterion**

  The original towing criterion in terms of the n's is expressed as:

  $$TWO = \frac{n_r n_l}{n^2} \left[\left| \frac{n_{1r}}{n_r} - \frac{n_{1l}}{n_l} \right| + \left| \frac{n_{0r}}{n_r} - \frac{n_{0l}}{n_l} \right| \right]^2 \ . \quad (4.48)$$

  Performing the same manipulations as above one arrives at

  $$TWO(p, \alpha, \beta) = (\alpha p + \overline{\beta} q)(\beta q + \overline{\alpha} p) \left[\left| \frac{\overline{\alpha} p}{\beta q + \overline{\alpha} p} - \frac{\alpha p}{\alpha p + \overline{\beta} q} \right| + \right.$$
  $$\left. + \left| \frac{\beta q}{\beta q + \overline{\alpha} p} - \frac{\overline{\beta} q}{\alpha p + \overline{\beta} q} \right| \right]^2 \ . \quad (4.49)$$

  $TWO(p, \alpha, \beta)$ is also *independent* of the predicted class labeling and its maximum value is $4pq$; therefore, we minimize $4pq - TWO(p, \alpha, \beta)$.

  The following two criteria are *not independent* of the predicted classification.

- **Probability-of-Error criterion:**

  Assuming the $r = \omega_1$, $l = \omega_0$ assignment and denoting by PE_1, PE_0 the class error rates ($PE_1 = n_{1l}/n_1 = 1 - \alpha$, $PE_0 = n_{0r}/n_0 = 1 - \beta$), we have:

  $$PE(p, \alpha, \beta) = pPE_1 + qPE_0 = p(1 - \alpha) + q(1 - \beta) \ . \quad (4.50)$$

Had we swapped the assigned labels we would obtain in general a different value: $PE(p, \alpha, \beta) = pPE_1 + qPE_0 = p\alpha + q\beta$.

- **Entropy-of-Error criterion:**

$$P_{10} \equiv P(E = 1) = p(1 - \alpha); \quad P_{01} \equiv P(E = -1) = q(1 - \beta) ; \qquad (4.51)$$

To these values we apply formula 4.46. Again, for swapped assignments one obtains in general a different value.

The three-dimensional functions expressed by formulas (4.46), (4.47), (4.49) and (4.50) are difficult to mathematically analyze and compare. It turns out to be more instructive to analyze these functions by considering two categories of configurations — $\alpha = \beta$ and $\alpha \neq \beta$ — and inspect the respective values.

Configurations with $\alpha = \beta$
For $\alpha = \beta$ configurations PE doesn't depend on p ($PE = 1 - \alpha$). All criteria depend on two parameters alone, p and PE, which facilitates their inspection as in Fig. 4.12. The behavior of TWO is similar to that of IG. We observe that SEE is concave, symmetric in p but asymmetric in PE! In conclusion, SEE displays, just as GI and IG, a desirable concave behavior *but more sensitive* to the PE value.

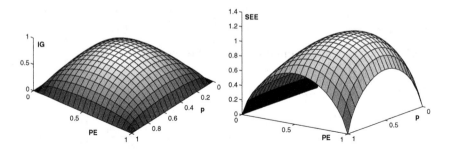

Fig. 4.12 IG and SEE for $\alpha = \beta$ (adapted from [152]).

Configurations with $\alpha \neq \beta$
Figure 4.13 shows the four criteria represented in a (α, β) grid for $p = 0.7$. We readily see that IG, TWO and SEE are concave functions whereas PE is a linear function. Figure 4.14 shows the error PMFs for four specifications of (α, β). Table 4.3 lists the respective criteria values. IG and TWO found cases 'a' and 'b' to be similar, which seems rather inadequate, whereas both PE and SEE found 'b' to be an improvement over 'a'. Cases 'b' and 'c' correspond to those analyzed in [33] to illustrate why PE should not be used. Note that IG and TWO select 'c' with a decrease of 16% and 11%, respectively; SEE also selects 'c' and with a more pronounced decrease: 22%. Therefore, for equal probability of error ($PE = 0.2$ in both cases), SEE also

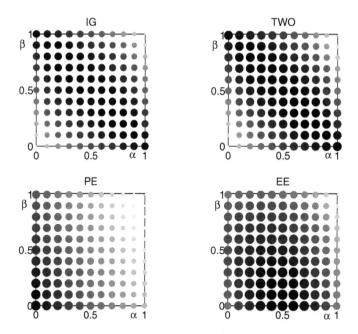

Fig. 4.13 The $p = 0.5$ setting. Darker tones and larger sizes correspond to higher values.

prefers purer nodes. Cases 'b' and 'd' constitute one further illustration of the sensitivity of SEE to PE, now in the $\alpha \neq \beta$ context. Similar conclusions can be drawn for other $p \in]0, 1[$ configurations.

Table 4.3 Criteria values for Fig. 4.14 configurations.

	α	β	IG	TWO	PE	SEE
a	0.49	0.01	0.49	0.68	0.75	1.04
b	0.80	0.80	0.50	0.64	0.20	0.64
c	0.60	1.00	0.42	0.57	0.20	0.50
d	0.70	0.89	0.50	0.64	0.21	0.63

To sum up, the empirical MEE (SEE) split behaves according to the theoretical properties regarding the class-conditional distribution overlapping, with the selection of empirical MEE split points generating a high interval-end hit rate for largely overlapped distributions. The SEE splitting function is a convenient concave function, as are the classic splitting functions (GI, IG, and TWO); however, SEE is more sensitive to the error rate, producing better splits in many simple situations. In Chap. 6 we will analyze how to use MEE splits in decision trees and what are the possible advantages.

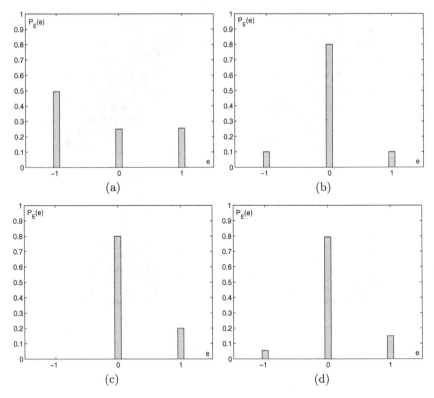

Fig. 4.14 Error PMFs for $p = 0.5$ with: a) $\alpha = 0.49$, $\beta = 0.01$; b) $\alpha = 0.8$, $\beta = 0.8$; c) $\alpha = 0.6$, $\beta = 1$; d) $\alpha = 0.7$, $\beta = 0.89$.

4.2 The Discrete-Output Perceptron

The perceptron with threshold activation function (also called step function), $z = \theta(y)$, implements a classifier function given by (see Sect. 3.3)

$$z = \theta(y) = \begin{cases} -1, & y = \mathbf{w}^T\mathbf{x} + w_0 \leq 0 \\ 1, & y = \mathbf{w}^T\mathbf{x} + w_0 > 0 \end{cases} , \qquad (4.52)$$

which is known as the McCulloch-Pitts model [156]. Rosenblatt [190] has proposed a training rule for this classifier which is proved to converge in a finite number of iterations provided the classes are linearly separable (the so-called perceptron convergence theorem). Here, we study how the SEE risk copes with the classifier problem (not restricted to linearly separable classes). We still use formula (4.2) but now the class error probabilities are given by

$$P_{-1} = P(Z = 1, T = -1) = P(\mathbf{w}^T\mathbf{x} + w_0 \geq 0, T = -1) = q(1 - F_{Y|-1}(0)),$$

$$(4.53)$$

$$P_1 = P(Z = -1, T = 1) = P(\mathbf{w}^T\mathbf{x} + w_0 \leq 0, T = 1) = pF_{Y|1}(0) , \qquad (4.54)$$

where $F_{Y|t}(0) = P(Y \leq 0|T = t)$ is the conditional distribution value at the origin of the univariate r.v. $Y = \mathbf{w}^T\mathbf{x} + w_0$.

A direct generalization of Theorem 4.2 can now be stated [212]:

Theorem 4.3. *In the two-class multivariate problem, if the optimal set of parameters given by $\mathbf{w}^* = [w_1^* \ \dots \ w_d^* \ w_0^*]^T$ of a separating hyperplane constitute a critical point of the error entropy, then the error probabilities of each class at $\mathbf{w}^*$ are equal.*

Proof. We start by noticing that the multivariate classification problem can be viewed has a univariate one using $u = \mathbf{w}^T\mathbf{x}$, the projection of $\mathbf{x}$ onto $\mathbf{w}$. From an initial input (overall) distribution represented by a density $f_X(\mathbf{x}) = qf_{X|-1}(\mathbf{x}) + pf_{X|1}(\mathbf{x})$ we get, on the projected space, the distribution of the projected data given by $f_U(u) = qf_{U|-1}(u) + pf_{U|1}(u)$. The parameter w_0 then works as a Stoller split: a data instance is classified as ω_1 if $u \geq w_0$ and as ω_{-1} otherwise. From Theorem 4.1, one can assert that $qf_{U|-1}(u) = pf_{U|1}(u)$ at $\mathbf{w}^*$.

We rewrite the error probabilities of each class as

$$P_{-1} = q(1 - F_{U|-1}(-w_0)) , \quad P_1 = pF_{U|1}(-w_0) , \qquad (4.55)$$

and compute

$$\frac{\partial P_{-1}}{\partial w_0} = -qf_{U|-1}(-w_0), \quad \frac{\partial P_1}{\partial w_0} = pf_{U|1}(-w_0) . \qquad (4.56)$$

From (4.2),

$$\frac{\partial H_S}{\partial P_t} = \ln\left(\frac{1 - P_{-1} - P_1}{P_t}\right), \qquad t \in \{-1, 1\} ,$$

the chain rule and the fact that $qf_{U|-1} = pf_{U|1}$ at $\mathbf{w}^*$ allows writting

$$\frac{\partial H_S}{\partial w_0}(\mathbf{w}^*) = 0 \Leftrightarrow \qquad (4.57)$$

$$\Leftrightarrow pf_{U|1}(w_0^*)\left(\ln\left(\frac{1 - P_{-1} - P_1}{P_{-1}}\right) - \ln\left(\frac{1 - P_{-1} - P_1}{P_1}\right)\right) = 0 \Leftrightarrow \quad (4.58)$$

$$\Leftrightarrow f_{U|1}(w_0^*) = 0 \vee P_{-1} = P_1 . \qquad (4.59)$$

Note that $f_{U|1}(w_0^*) = 0$ iff the classes have distributions with disjoint supports (they are separable). But in this case $P_{-1} = P_1 = 0$. Thus, in both cases $P_{-1} = P_1$ is a necessary condition. $\qquad \square$

We remark that the proof has only analyzed $\frac{\partial H_S}{\partial w_0}$ because the other derivatives (and consequently the complete gradient ∇H_S) are rather intricate. Thus, equality of class error probabilities is just a necessary condition. The following example illustrates this result.

Example 4.5. Consider the perceptron implementing the family of lines $w_1 x_1 + w_2 x_2 + w_0 = 0$ to discriminate between two bivariate Gaussian classes. First, let $\mu_{\pm 1} = [\pm 2\ 0]$ and $\Sigma_1 = \Sigma_{-1} = \mathbf{I}$. The optimal solution is given (as a function of p) by the vertical line with equation

$$x_1 = \frac{1}{4} \ln \left(\frac{1-p}{p} \right) .$$

The optimal set of parameters must satisfy $w_2^* = 0$ and $w_0^* = -\frac{w_1^*}{40} \ln \left(\frac{1-p}{p} \right)$. Additionally, w_1^* must be positive to give the correct class orientation. One can then numerically determine that $\nabla H_S(\mathbf{w}^*) = \mathbf{0}$ only if $p = 1/2$, which corresponds to the class setting with equal class error probabilities.

If we now assume $p = 1/2$ and $\Sigma_1 = \left[\begin{smallmatrix} 2 & 0 \\ 0 & 1 \end{smallmatrix} \right]$, the optimal solution is

$$x_1 = -6 + \sqrt{32 + 2\ln(2)} .$$

The error probabilities are unequal, $P_{-1} \approx 0.019$ and $P_1 \approx 0.029$, and

$$\nabla H_S(w_1^*, 0, w_1^*(-6 + \sqrt{32 + 2\ln(2)})) \neq \mathbf{0} . \tag{4.60}$$

More precisely, $\frac{\partial H_S}{\partial w_1} < 0$, $\frac{\partial H_S}{\partial w_2} = 0$ and $\frac{\partial H_S}{\partial w_0} > 0$ at the possible optimal solutions. Therefore, the optimal solution is not a critical point of the error entropy.   □

The above example indicates that it suffices from now on to analyze the case of bivariate Gaussian class distributions to get a picture of the discrete MEE (SEE) behavior regarding the optimality issue. Recall from Sect. 3.3.1 that Gaussianity is preserved under linear transformations. Therefore, if the classes have means μ_t and covariances Σ_t for $t \in \{-1, 1\}$, it is straightforward to obtain

$$F_{U|t}(0) = \Phi \left(-\frac{\mathbf{w}^T \mu_t + w_0}{\sqrt{\mathbf{w}^T \Sigma_t \mathbf{w}}} \right) . \tag{4.61}$$

For equal priors one gets

$$P_{-1} = \frac{1}{2} \left(1 - \Phi \left(-\frac{\mathbf{w}^T \mu_{-1} + w_0}{\sqrt{\mathbf{w}^T \Sigma_{-1} \mathbf{w}}} \right) \right); \quad P_1 = \frac{1}{2} \Phi \left(-\frac{\mathbf{w}^T \mu_1 + w_0}{\sqrt{\mathbf{w}^T \Sigma_1 \mathbf{w}}} \right) . \tag{4.62}$$

Unfortunately these expressions imply a rather intricate entropy formula and of the corresponding derivatives. Let us consider spherical distributions with $\Sigma_{-1} = \Sigma_1 = \mathbf{I}$, to obtain a linear (optimal) solution and, in order to simplify the above expressions, assume that the centers (means) of the classes lie in the horizontal axis and are symmetric with respect to the origin (every possible class configuration can be reduced to this case through shifts and rotations). The optimal decision boundary is the vertical line $x_1 = 0$ with the optimal set of parameters $[w_1^* \ w_2^* \ w_0^*] = [c \ 0 \ 0]$ with $c \in \mathbb{R}^+$ to give the correct class orientation. Let us analyze SEE solutions of the form $\bar{\mathbf{w}} = [w_1 \ 0 \ 0]$ with $w_1 \neq 0$. We first note that $H_S(\bar{\mathbf{w}})$ is piecewise constant

$$H_S(\bar{\mathbf{w}}) = \begin{cases} c_1, & w_1 > 0 \\ c_2, & w_1 < 0 \end{cases} \tag{4.63}$$

where $c_1 \rightarrow 0$ (100 % correct classification) or $c_1 \rightarrow \ln(0.5)$ (swapped labels) for an increasing distance between the classes. Also, $\nabla H_S(\bar{\mathbf{w}}) = 0$, that is, the vectors $\bar{\mathbf{w}}$ are critical points of SEE. Considering two particular class configurations the Hessian $\nabla^2 H_S(\bar{\mathbf{w}})$ will give us insight about the nature of $\bar{\mathbf{w}}$ for $w_1 > 0$.

1. Distant classes: $\mu_{\pm 1} = [\pm 5 \ 0]^T$
   The Hessian matrix $\nabla^2 H_S$ at $\bar{\mathbf{w}}$ is given by

$$\nabla^2 H_S(\bar{\mathbf{w}}) = \begin{bmatrix} 0 & 0 & 0 \\ 0 & \frac{0.4809}{w_1^2} & 0 \\ 0 & 0 & \frac{0.3527}{w_1^2} \end{bmatrix},$$

which is a positive semi-definite matrix. The nature of $\bar{\mathbf{w}}$ cannot be evaluated directly due to the singularity of the Hessian. Using the Taylor expansion of H_S to analyze its behavior in a neighborhood of $\bar{\mathbf{w}}$ one gets

$$H_S(\bar{\mathbf{w}} + \mathbf{h}) - H_S(\bar{\mathbf{w}}) \approx \mathbf{h}^T \nabla^2 H_S(\bar{\mathbf{w}}) \mathbf{h} . \tag{4.64}$$

where $\mathbf{h} = [h_1 \ h_2 \ h_3]^T$ is a small increment on $\bar{\mathbf{w}}$.
It is easily seen that there are increments $\bar{\mathbf{h}} = [h_1 \ 0 \ 0]$ such that $\bar{\mathbf{h}}^T \nabla^2 H_S(\bar{\mathbf{w}}) \bar{\mathbf{h}} = 0$. But $\bar{\mathbf{w}} + \bar{\mathbf{h}}$ belongs to the positive w_1 axis where H_S is constant. Along any other $\mathbf{h}$ directions, the quadratic form is positive. This means that $\bar{\mathbf{w}}$, or more precisely, the whole positive w_1 axis, is in fact an entropy minimum.
2. Close classes: $\mu_{\pm 1} = [\pm 0.5 \ 0]^T$

The Hessian now becomes

$$\nabla^2 H_S(\bar{\mathbf{w}}) = \begin{bmatrix} 0 & 0 & 0 \\ 0 & \frac{0.2641}{w_1^2} & 0 \\ 0 & 0 & \frac{-0.1377}{w_1^2} \end{bmatrix} . \tag{4.65}$$

This matrix is indefinite (it has positive and negative eigenvalues). This means that there are directions such that $\bar{\mathbf{w}}$ is a minimum and directions such that $\bar{\mathbf{w}}$ is a maximum (and, as discussed above, directions where H_S remains constant). So, $\bar{\mathbf{w}}$ is a saddle point.   □

The above analysis again shows that SEE has a different behavior depending on the distance between the classes. A graphical insight of this behavior can be found in [212]. The following Example shows how to determine the t_{value} for Gaussian classes.

Example 4.6. The eigenvalues of the Hessian can be used to compute the minimum-to-saddle turn-about value for bivariate Gaussian classes. As $\nabla^2 H_S$ $(\bar{\mathbf{w}})$ is always a diagonal matrix with one zero entry, one positive entry, and a third entry that changes sign as the classes get closer, corresponding to the eigenvalues of the matrix, one can determine the minimum distance yielding a minimum of H_S at $[w_1 \; 0 \; 0]^T$ for $w_1 \neq 0$, by inspecting when the sign-changing eigenvalue changes of sign. Setting $\Sigma_1 = \Sigma_2 = \sigma^2 \mathbf{I}$ and the class means symmetrical about the origin and at the horizontal axis, it is possible to derive that the third eigenvalue is positive if the following expression is positive:

$$\sqrt{2\pi}d(1 - \Phi(d))\ln\left(\frac{2\Phi(d)}{1 - \Phi(d)}\right) - e^{-\frac{d^2}{2}}, \tag{4.66}$$

where d is a normalized half distance between the classes. The turn-about value is $d = 0.7026$, which corresponds to a normalized distance between the classes of 1.4052. This is precisely the same value found for the Stoller split problem in Example 4.3.   □

Chapter 5
EE-Inspired Risks

In the previous chapters the behavior of classifiers trained to minimize error-entropy risks, for both discrete and continuous errors, was analyzed. The rationale behind the use of these risks is the fact that entropy is a PDF concentration measure — higher concentration implies lower entropy —, and in addition (recalling what was said in Sect. 2.3.1) minimum entropy is attained for Dirac-δ combs (including a single one). Ideally, in supervised classification, one would like to drive the learning process such that the final distribution of the error variable is a Dirac-δ centered at the origin. In rigor, this would only happen for completely separable classes when dealing with the discrete error case or with infinitely distant classes when dealing with the continuous error case with the whole real line as support. For practical problems we will content ourselves in achieving the highest possible error concentration at the origin, namely by driving the classifier training process to an error PDF (or PMF) with the highest possible value at the origin. Formally, the process may be expressed as

$$\text{Find } \bar{\mathbf{w}} \text{ s.t. } \bar{\mathbf{w}} = \arg\max_{\mathbf{w} \in W} f_E(\mathbf{0}; \mathbf{w}) \text{ for continuous errors} \qquad (5.1)$$

or

$$\bar{\mathbf{w}} = \arg\max_{\mathbf{w} \in W} P_E(\mathbf{0}; \mathbf{w}) \text{ for discrete errors} \qquad (5.2)$$

where $\mathbf{w}$ is the parameter vector of the classifier and $\bar{\mathbf{w}}$ its optimal value in the Zero-Error Density (Probability) Maximization (ZED(P)M) sense (first proposed in the 2005 work [213]). In the following we will concentrate on the continuous error case and explore this approach by deriving the Zero-Error Density (ZED) risk and further evolve to a generalized risk capable of emulating a whole family of risks [218]. As before, we focus on the two-class case.

J. Marques de Sá et al.: Minimum Error Entropy Classification, SCI 420, pp. 121–137.
springerlink.com   © Springer-Verlag Berlin Heidelberg 2013

5.1 Zero-Error Density Risk

The ZED risk, R_{ZED}, is obtained by simply evaluating the error PMF or PDF at the origin, that is,

$$R_{ZED} \equiv R_{ZED}(E) = \begin{cases} P_E(0), & \text{discrete } E \\ f_E(0) = qf_{E|-1}(0) + pf_{E|1}(0), & \text{continuous } E \end{cases} .$$

$$(5.3)$$

Note that we are omitting for notation simplicity the dependency of R_{ZED} on the parameter vector $\mathbf{w}$. We emphasize that we are looking for a maximum of $P_E(0)$ or $f_E(0)$ as a funtion of $\mathbf{w}$, that is, looking for the classifier's parameters that provide the highest possible error PDF "concentration" at the origin. $P_E(0)$ or $f_E(0)$ may eventually be local minimum values of $P_E(e)$ or $f_E(e)$, respectively. In the following examples we analyze the case of a simple machine: the data splitter.

5.1.1 Motivational Examples

We present two examples, the first one illustrating the ZEPM approach with a successful outcome; the second one, illustrates the ZEDM approach providing an unsuccessful outcome.

Example 5.1. Consider the discrete data splitter introduced in Chap. 4 with (single) parameter $w = x'$. From formulas (4.1) one may write

$$P_E(0) = 1 - P_{-1} - P_1 . \tag{5.4}$$

Now,

$$\frac{dP_E(0)}{dw}(w) = 0 \iff qf_{X|-1}(w) = pf_{X|1}(w) , \tag{5.5}$$

that is, the critical points are located at the intersections of the posterior class-conditional PDFs. For continuous $f_{X|t}$, this is also the location of w^* (Theorem 4.1), the min P_e solution. Moreover,

$$\frac{d^2 P_E(0)}{dw^2}(w) = q\frac{df_{X|-1}}{dw}(w) - p\frac{df_{X|1}}{dw}(w) , \tag{5.6}$$

and for a very large family of two-class problems (e.g., the mutually symmetric distributions described in 4.1.2), the derivatives have oposite signs at w^* with the ω_1 "right-hand" class having $\frac{df_{X|1}}{dw}(w^*) > 0$. Thus, $P_E(0)|_{w^*}$ is a maximum of $P_E(0)$.   □

Example 5.2. Consider now the continuous-output data splitter $y = \tanh(x - w_0)$ introduced in Sect. 3.5. For Gaussian input class-conditional PDFs the

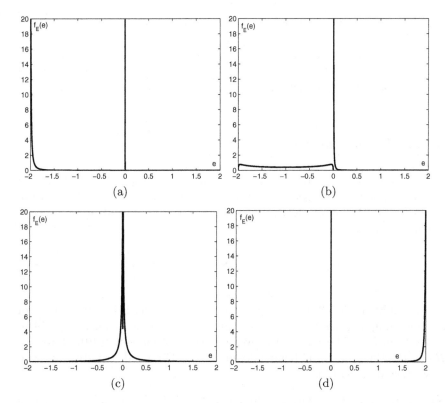

Fig. 5.1 $f_E(e)$ for the continuous data splitter $\tanh(x-w_0)$ in a two-class Gaussian problem with $\mu_1 = -\mu_{-1} = 2$ and $\sigma_1 = \sigma_{-1} = 1$: a) $w_0 = -5$; b) $w_0 = -2$; c) $w_0 = 0 = w_0^*$; d) $w_0 = 5$.

error PDF is given by

$$f_E(e) = q f_{E|-1}(e) + p f_{E|1}(e) , \tag{5.7}$$

with (see Sect. 3.3.2.1)

$$f_{E|t}(e) = \frac{\exp\left(-\frac{1}{2}\left(\frac{\operatorname{atanh}(t-e)-(\mu_t-w_0)}{\sigma_t}\right)^2\right)}{\sqrt{2\pi}\sigma_t e(2t-e)} \mathbb{1}_{]t-1,t+1[}(e) . \tag{5.8}$$

We note that $f_E(0)$ is not defined but $\lim_{e\to 0} f_E(e)$ exists and is zero. We now analyze what happens when w_0 varies. Figure 5.1 shows $f_E(e)$ for different values of w_0 and the following settings of the two-class problem: $\mu_1 = -\mu_{-1} = 2$ and $\sigma_1 = \sigma_{-1} = 1$. For $w_0 \to -\infty$ (of which Fig. 5.1a is an example), $f_{E|-1}$ and $f_{E|1}$ converge to Dirac-δ centered at $e = -2$ and $e = 0$, respectively. A similar behavior is found for $w_0 \to +\infty$ where $f_{E|-1}$ and $f_{E|1}$ converge to a

Dirac-δ centered at $e = 0$ and $e = 2$, respectively (see Fig. 5.1d). Now, for $w_0 = 0$ (Fig. 5.1c), the min P_e solution, one has a finite value for $f_E(e)$ in a neighborhood of $e = 0$, always lower than the maximum value attained for other values of w_0. This unsuccessful outcome for the theoretical ZEDM will not hold for the empirical version.   □

5.1.2 Empirical ZED

From now on we exclusively focus the continuous error case. The practical application of ZED necessarily implies the estimation of the error PDF at the origin, $f_E(0)$. As before, KDE is used as in formula (3.2) allowing to write the empirical risk as

$$\hat{R}_{ZED} = \hat{f}_E(0) = \frac{1}{n} \sum_{i=1}^{n} \frac{1}{h} K \left(\frac{0 - e_i}{h} \right) . \tag{5.9}$$

We saw in Example 5.2 that $f_E(0)$ is not defined and this is in fact a feature of sigmoid-equipped classifiers (like the logistic or tanh functions). This problem is "fixed" in $\hat{R}_{ZED}$ because of the oversmoothed estimate of $f_E(0)$. Moreover, $\hat{R}_{ZED}$ has the property of being maximum for a Dirac-δ distribution at $e = 0$ as we now demonstrate.

Theorem 5.1. *The empirical estimate of the ZED risk, $\hat{R}_{ZED}$, has a global minimum for a Dirac-δ distribution of the errors centered at the origin.*

Proof. Let $\mathbf{e} = [e_1 \ldots e_n]^T$ be the vector that collects all the errors produced by the learning process. $\hat{R}_{ZED}$ can be seen as a function of $\mathbf{e}$ and its gradient and Hessian can be computed for a Dirac-δ of the errors at the origin, that is, at $\mathbf{e} = \mathbf{0}$. Now,

$$\frac{\partial \hat{f}_E(0)}{\partial e_k} = -\frac{1}{nh^2} K' \left(-\frac{e_k}{h} \right) \tag{5.10}$$

and therefore

$$\left. \frac{\partial \hat{f}_E(0)}{\partial e_k} \right|_{\mathbf{e}=0} = 0 \ \Leftrightarrow \ K'(0) = 0 . \tag{5.11}$$

Moreover,

$$\frac{\partial^2 \hat{f}_E(0)}{\partial e_k^2} = \frac{1}{nh^3} K'' \left(-\frac{e_k}{h} \right) , \tag{5.12}$$

$$\frac{\partial^2 \hat{f}_E(0)}{\partial e_j \partial e_k} = 0 \qquad \forall j \neq k . \tag{5.13}$$

Thus, the Hessian matrix at $\mathbf{e} = \mathbf{0}$ is of the form $\frac{1}{nh^3}K''(0)\mathbf{I}$ which means that it has an unique eigenvalue $\lambda = \frac{1}{nh^3}K''(0)$ with multiplicity n. The Hessian is positive definite, and thus $\hat{R}_{ZED}$ has a maximum at $\mathbf{e} = \mathbf{0}$, if $K''(0) < 0$. We can also show that the maximum is global. First note that

$$\hat{f}_E(0)\Big|_{\mathbf{e}=\mathbf{0}} \geq \hat{f}_E(0)\Big|_{\mathbf{e}} \quad \Leftrightarrow \quad nK(0) \geq \sum_{i=1}^{n} K\left(-\frac{e_i}{h}\right) . \tag{5.14}$$

If the kernel function is unimodal with mode at the origin, then

$$\sum_{i=1}^{n} K\left(-\frac{e_i}{h}\right) \leq n \max_i K\left(-\frac{e_i}{h}\right) \leq nK(0) . \tag{5.15}$$

In short, if $K'(0) = 0$, $K''(0) < 0$ and K is unimodal with mode at the origin, then $\hat{R}_{ZED}$ attains its global maximum at $\mathbf{e} = \mathbf{0}$. $\qquad\square$

The Gaussian kernel satisfies the three properties of Theorem 5.1, and is thus to be preferred for practical applications. Hence,

$$\hat{R}_{ZED} = \hat{f}_E(0) = \frac{1}{n}\sum_{i=1}^{n}\frac{1}{h}G\left(\frac{e_i}{h}\right) . \tag{5.16}$$

The kernel bandwidth (or smoothing parameter) h also plays an important role in the success of this approach as illustrated in the following example.

Example 5.3. Consider the two-class setting of Example 5.2. Let us analyze the influence of h in estimating $\hat{R}_{ZED}$. Figure 5.2 shows $\hat{f}_E(0)$ plotted as a function of w_0 for Gaussian generated input data (10 000 instances for each class). Recall from Example 5.2 that $f_E(0)$ is not defined, but here the KDE produces a smoothed $\hat{f}_E(0)$ value. For small values of h ($h < h_{IMSE} = 0.27$;

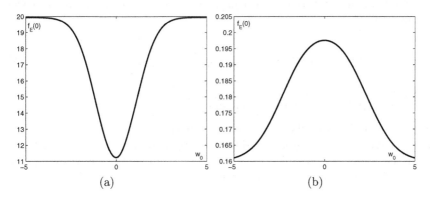

Fig. 5.2 $\hat{f}_E(0)$ plotted as a function of w_0 for the two-class setting of Example 5.2 with two different values of h: a) $h = 0.01$; b) $h = 2$.

see Sect. E.2) we are closely estimating $f_E(0)$ and $\hat{R}_{ZED}$ exhibits what was observed in Example 5.2: $f_E(0)\big|_{w_0^*}$ has a lower value than $\hat{f}_E(0)\big|_{w_0}$ for any other off-optimal w_0 values. In particular, $\hat{f}_E(0)\big|_{w_0}$ increases for $w_0 \to \pm\infty$. On the other hand, using a large value for h, $\hat{f}_E(0)$ attains its maximum at $w_0 = 0$, corresponding to the min P_e solution. Note that we are not interested in an accurate estimate of the error PDF but only to "force" it to have a large value at the origin. By using a large h value we are computing an oversmothed KDE (fat estimation) that suffices to mask off-optimal solutions and highlight the optimal ones.   □

5.1.2.1 ZED Gradient Ascent

Gradient ascent optimization can be applied to maximize $\hat{R}_{ZED}$ as follows:

Algorithm 5.1 – ZED Gradient ascent algorithm

1. Choose a random initial parameter vector, $\mathbf{w}$ (with components w_k).
2. Compute the error PDF estimate at the origin using the classifier outputs $y_i = \varphi(\mathbf{x}_i; \mathbf{w})$ at the n available instances.
3. Compute the partial derivatives of $\hat{R}_{ZED}$ with respect to the parameters:

$$\frac{\partial \hat{f}_E(0)}{\partial w_k} = -\frac{1}{nh^2} \sum_{i=1}^{n} G'\left(-\frac{e_i}{h}\right) \frac{\partial e_i}{\partial w_k} =$$

$$= -\frac{1}{nh^3} \sum_{i=1}^{n} e_i G\left(-\frac{e_i}{h}\right) \frac{\partial e_i}{\partial w_k} \tag{5.17}$$

where $\partial e_i / \partial w_k$ depends on the classifier architecture.

4. Update at each iteration, m, the parameters $w_k^{(m)}$ using a η amount (learning rate) of the gradient:

$$w_k^{(m)} = w_k^{(m-1)} + \eta \left.\frac{\partial \hat{f}_E(0)}{\partial w_k}\right|_{w_k^{(m-1)}}. \tag{5.18}$$

5. Go to step 2, if some stopping criterion is not met.

It is interesting to note that, in constrast with the EE risks, this algorithm has a $O(n)$ complexity similar to the MMSE or MCE gradient descent algorithms.

In the following examples we apply Algorithm 5.1 to the training of a perceptron solving a two-class problem.

Example 5.4. Consider two Gaussian distributed class conditional PDFs with

$$\mu_{-1} = [0 \ 0]^T; \mu_1 = [3 \ 0]^T; \Sigma_{-1} = \Sigma_1 = \mathbf{I}. \tag{5.19}$$

Independent training and test sets are generated with $n = 200$ instances (100 per class) and $\hat{R}_{ZED}$ is used with $h = 1$ and $h = 3$. An initial $\eta = 0.001$ is used. Figs. 5.3 and 5.4 show the final converged solution after 80 epochs of training: the final decision border and error PDF estimate; $\hat{R}_{ZED}$ and the training and test misclassification rates along the training process. We see that an increased h provokes a need of more initial epochs till $\hat{R}_{ZED}$ starts to increase significantly. This is explained by looking to formula (5.17): a higher h implies a lower $\partial \hat{f}_E(0)/\partial w_k$ and consequently smaller gradient ascent steps (the adaptive η then compensates the influence of the higher h). On the other hand, a better generalization is obtained with $h = 3$. In fact, when using a lower h the perceptron provides a better discrimination on the training set (look to the decision borders of both figures) but with an increased test set error. The use of a higher h provides an oversmoothed estimate of $\hat{R}_{ZED}$, masking local off-optimal solutions.

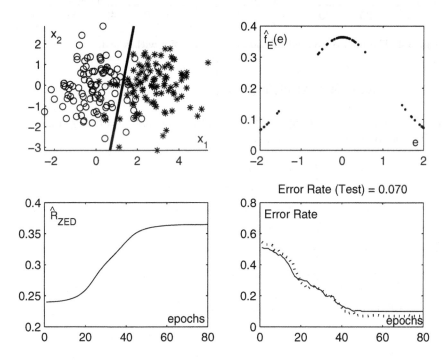

Fig. 5.3 The final converged solution of Example 5.4 with $h = 1$.

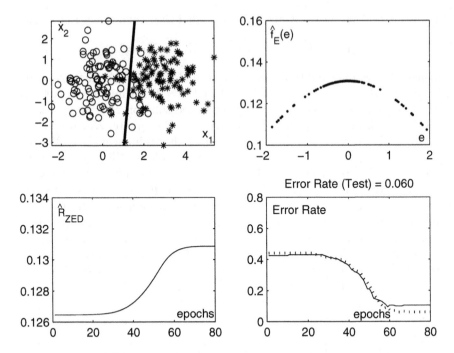

Fig. 5.4 The final converged solution of Example 5.4 with $h = 3$.

5.1.2.2   Comparing MSE, CE, and ZED

Using formulas (2.5), (2.28) and (5.16) and noting that $e_i = t_i - y_i$, one may write, for some classifier parameter w that

$$\frac{\partial \hat{R}_{MSE}}{\partial w} = -\frac{1}{n} \sum_{i=1}^{n} e_i \frac{\partial y_i}{\partial w}, \tag{5.20}$$

$$\frac{\partial \hat{R}_{CE}}{\partial w} = -\sum_{i=1}^{n} \frac{e_i}{y_i(1 - y_i)} \frac{\partial y_i}{\partial w}, \tag{5.21}$$

$$\frac{\partial \hat{R}_{ZED}}{\partial w} = \frac{1}{nh^3} \sum_{i=1}^{n} e_i G\left(-\frac{e_i}{h}\right) \frac{\partial y_i}{\partial w}. \tag{5.22}$$

Define the following weight functions

$$\psi_{MSE}(e) = e, \tag{5.23}$$

$$\psi_{CE}(y) = \frac{e}{y(1 - y)} = \frac{t - y}{y(1 - y)}, \qquad t \in \{0, 1\}, \tag{5.24}$$

$$\psi_{ZED}(e) = e\, G\left(-\frac{e}{h}\right). \tag{5.25}$$

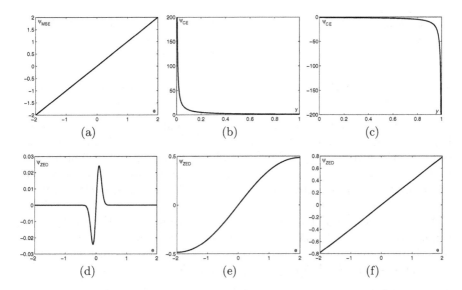

Fig. 5.5 Weight functions: a) ψ_{MSE}; b) ψ_{CE} for $t = 1$; c) ψ_{CE} for $t = 0$; d) ψ_{ZED} for $h = 0.1$; e) ψ_{ZED} for $h = 1$; f) ψ_{ZED} for $h = 10$

We may then write formulas (5.20) to (5.22) as

$$\frac{\partial R}{\partial w} = k \sum \psi(e_i) \frac{\partial y_i}{\partial w} \, .$$

Here we omit the constants $1/n$ and $1/nh^3$ from ψ_{MSE} and ψ_{ZED}, respectively. This is unimportant from the point of view of optimization as one could always multiply $\hat{R}_{MSE}$ and $\hat{R}_{ZED}$ by n and nh^3 respectively without affecting their extrema. As discussed in [214] this only affects the behavior of the learning process by increasing the number of necessary epochs to converge.

Figure 5.5 presents a comparison of the behavior of the weight functions. From Fig. 5.5a we see that ψ_{MSE} is linear such that each error contributes with a weight equal to its own value. Thus, larger errors are more penalized contributing with a larger weight for the whole gradient. On the other hand, ψ_{CE} confers even larger weights to larger errors. As Figs. 5.5b and 5.5c show this weight assignment follows a hyperbolic-type rule (in contrast with the linear rule of ψ_{MSE}). Now, for ψ_{ZED} one may distinguish three basic behaviors:

1. If h is small, as in Fig. 5.5d, $\psi_{ZED}(e) \approx 0$ for a large sub-domain of the variable e. This may cause difficulties for the learning process to converge (or even start at all). In fact, it is common procedure to randomly initialize the classifier's parameters with values close to zero, producing errors around $e = -1$ and $e = 1$. In this case, the learning process would not converge.

2. For moderate h (say $h \approx 2$ as in Fig. 5.5e), ψ_{ZED} shows a sigmoidal-type shape and, as in ψ_{MSE} and ψ_{CE}, larger errors contribute to larger weights. Note, however, the contrast with ψ_{CE}: for larger errors ψ_{CE} "accelerates" the weight value while ψ_{ZED} "decelerates".

3. For larger values of h, ψ_{ZED} behaves like ψ_{MSE}, as illustrated in Fig. 5.5f. In fact, $\lim_{h \to +\infty} \psi_{ZED} = \psi_{MSE}$.

Despite the disadvantage of $\hat{R}_{ZED}$ over $\hat{R}_{MSE}$ and $\hat{R}_{CE}$ in having to set h, it is important to emphasize that we are not concerned in obtaining a good estimate of $\hat{f}_E(0)$ but only to force it to be as high as possible. This means that we can set some moderately high value for h with the advantage of adapting it, and thus controlling how ψ_{ZED} behaves, for each classification problem at hand.

Moreover, the second basic behavior above suggests that the "decelerated" caracteristic of ψ_{ZED} enables a reduced sensitivity of $\hat{R}_{ZED}$ to outliers (the sensitivity degree controlled by h) when compared to the other alternative risks. This is illustrated in the following example.

Example 5.5. Consider discriminating two classes with bivariate input data $\mathbf{x} = [x_1 \ x_2]^T$, with circular uniform distribution (see Example 3.8 in Sect. 3.3.1) and the following parameters:

$$\mu_{-1} = [0 \ 0]^T, \quad \mu_1 = [1.1 \ 0]^T, \quad r_{-1} = r_1 = 1 . \qquad (5.26)$$

By symmetry the theoretically optimal linear discriminant is orthogonal to x_1 at the decision threshold $d = -w_0/w_1 = 0.55$ and with $\min P_e = 0.1684$.

Suppose that a training set from the said distributions with n instances per class was available, which for whatever reason was "contaminated" by the addition to class ω_{-1} of n_0 instances, $n_0 \ll n$, with uniform distribution in $]1, 1 + l]$ along x_1. Figure 5.6 shows an example of such dataset with $n = 200$ instances per class and $n_0 = 10$ outliers uniformly distributed in $]1, 1 + l]$ with $l = 0.2$ (solid circles extending beyond $x_1 = 1$). Also shown is a linear discriminant adjusted by an $\hat{R}_{ZED}$ perceptron trained with $h = 1$ (fat estimation of the error PDF) during 80 epochs with $\eta = 0.001$.

In order to investigate the influence of the n_0 outliers in the determination of the decision threshold d, we proceed as follows: we repeat n_{exp} times the experiment of randomly generating datasets with $2n + n_0$ instances ($n + n_0$ instances for class ω_{-1}, and n instances for class ω_1) and train $\hat{R}_{ZED}$ and $\hat{R}_{MSE}$ perceptrons always with the above settings (80 epochs, $\eta = 0.001$, $h = 1$). We do this for several values of l, governing the spread of the outliers.

Figure 5.7 shows averages of $d \pm std(d)$ in terms of l, obtained in $n_{exp} = 500$ experiments, for datasets with $n = 200$ instances per class and two values of n_0: $n_0 = 10$ (Fig. 5.7a) and $n_0 = 20$ (Fig. 5.7b). The value $l = 1$ corresponds to the no outlier case. The experimental results shown in Fig. 5.7 clearly indicate that the average d for the $\hat{R}_{ZED}$ perceptron (thick dashed line) is

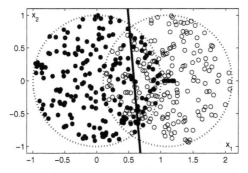

Fig. 5.6 A linear discriminant (solid line) obtained by training an $\hat{R}_{ZED}$ perceptron for a two-class dataset with outliers.

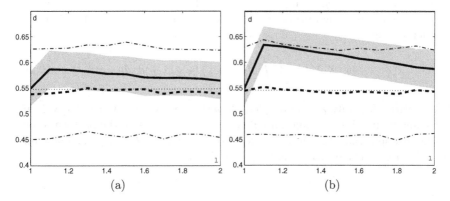

Fig. 5.7 Average d for the $\hat{R}_{ZED}$ perceptron (thick dashed line) and the $\hat{R}_{MSE}$ perceptron (thick solid line) for two values of n_0: a) $n_0 = 10$; b) $n_0 = 20$. The shadowed region represents the $\pm std(d)$ interval for the $\hat{R}_{MSE}$ perceptron; the dash-dot lines are the $\pm std(d)$ interval limits for the $\hat{R}_{ZED}$ perceptron.

always close to the theoretical, outlier free, $d = 0.55$, whereas the average d for the $\hat{R}_{MSE}$ perceptron (thick solid line) deviates considerably from the theoretical value. Briefly, the $\hat{R}_{ZED}$ solution is, on average, quite insensitive to the presence of outliers, whereas the same cannot be said of the average $\hat{R}_{MSE}$ solution. Note that the deviations between the $\hat{R}_{ZED}$ and $\hat{R}_{MSE}$ solutions are larger for small l, when an outlier probability peak stands near the circular uniform border, and therefore with close influence on decision border adjustment; large values of l correspond to tenuous low-probability tails with less influence on decision border adjustment.   □

5.1.2.3 ZED versus Correntropy

The ZED risk functional can also be derived from the framework of the *cross-correntropy* or simply *correntropy* function, a generalized correlation function first proposed in the 2006 work [196] (the name was chosen to reflect the connection of the proposed correlation measure to Rényi's quadratic entropy estimator). Later, a more general form of correntropy was proposed [141,174], as a generalized similarity measure between two arbitrarily scalar random variables X and Y defined as

$$v(X,Y) = \int \int K(x,y)f_{X,Y}(x,y)dxdy \; . \tag{5.27}$$

Here, $K(x,y)$ is any continuous positive definite kernel with finite maximal value. If $K(x,y) = xy$, then the conventional cross-correlation is obtained as a special case, but the authors concentrate on the special case of a Gaussian kernel, giving

$$v_h(X,Y) = \int \int G_h(x-y)f_{X,Y}(x,y)dxdy \; . \tag{5.28}$$

Correntropy is positive and bounded (in particular, $0 < v_h(X,Y) \le 1/\sqrt{2\pi h^2}$), reaching its maximum if and only if $X = Y$. To put this on the pattern recognition framework, consider as before the error variable $E = T - Y$. Then, one can define the correntropy between T and Y as

$$v_h(T,Y) = \mathbb{E}_{X,Y}\{G_h(T-Y)\} = \mathbb{E}_E\{G_h(E)\} = \int G_h(e)f_E(e)de. \tag{5.29}$$

An important property now appears [141, 174]. If $h \to 0$ then $v_h(T,Y)$ amounts to $f_E(0)$, that is,

$$\lim_{h \to 0} v_h(T,Y) = f_E(0) \; . \tag{5.30}$$

So, maximizing the correntropy (coined MCC in [141]) between the target and output variables is equivalent to maximize the error density at the origin, provided a sufficiently small h is considered. This is the same idea as the ZEDM principle with risk functional given by R_{ZED} as in formula (5.3).

The empirical version of correntropy is obtained by noticing that $v_h(T,Y)$ is an expected value, giving the following sample estimate

$$\hat{v}_h(T,Y) = \frac{1}{n}\sum_{i=1}^{n} G_h(e_i) = \hat{f}_E(0) \; , \tag{5.31}$$

which is precisely the empirical ZED risk of formula (5.16) with Gaussian kernel. This estimator has some good properties as shown in [141, 174]: if

$n \rightarrow \infty$, $\hat{v}_h(T,Y)$ is a consistent (in the mean square sense) estimator of $v_h(T,Y)$; moreover, if $nh \rightarrow \infty$ and $h \rightarrow 0$, $\hat{v}_h(T,Y)$ is an asymptotically unbiased and consistent estimator of $f_E(0)$.

Another interesting property is that correntropy is closely related to Huber's M-estimation [106] via the so-called *correntropy induced metric* (CIM):

$$CIM(T,Y) = \sqrt{v(0,0) - v(T,Y)} \ . \tag{5.32}$$

This metric applies different evaluations of distance depending on how far the samples are: it goes from a L_2 metric, when samples are close, to a L_0 metric when samples are distant [141, 174]. This close-to-distant relation is controlled by the kernel bandwidth h. Thus, correntropy and by consequence the ZED risk, can be seen as robust risk functionals with a reduced sensitivity to outliers. This theoretical results support what was discussed at the end of the previous section and illustrated in Example 5.5.

For additional properties of correntropy and practical applications consult, for example, [141, 174, 196, 118, 70].

5.2 A Parameterized Risk

In Sect. 5.1.2.2 we compared the gradients of MSE, CE and ZED empirical risks to understand their behavior in iteractive training, namely when using the gradient descent (ascent for ZED) algorithm. We argued that having a free parameter (the kernel bandwidth h) gives us an extra flexibility allowing to adapt the risk functional to the particular classification problem at hand. Moreover, we showed that for increasingly h in $\hat{R}_{ZED}$ we recover the MSE gradient behavior, getting a 2-in-1 risk functional. Would it also be possible to incorporate the CE risk type behavior in a general parameterized family of risk functionals? The answer has been shown to be afirmative [218] as we now present.

5.2.1 The Exponential Risk Functional

We define the (empirical) *Exponential* risk functional as

$$\hat{R}_{EXP} = \sum_{i=1}^{n} \tau \exp\left(\frac{e_i^2}{\tau}\right) \tag{5.33}$$

for a real parameter $\tau \neq 0$. We start by noticing that for $\tau = -2h^2$ we have

$$\hat{R}_{ZED} = \frac{1}{\tau\sqrt{2\pi}hn}\hat{R}_{EXP} \ , \tag{5.34}$$

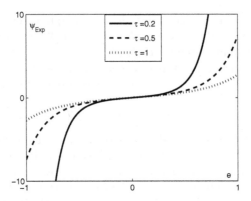

Fig. 5.8 Plot of ψ_{Exp} for different $\tau > 0$.

that is, $\hat{R}_{ZED}$ can be seen as a special case of $\hat{R}_{EXP}$ if we consider $\tau < 0$. Thus, for $\tau \to -\infty$, $\hat{R}_{EXP}$ gradient behaves like the MSE counterpart. Since $e_i = t_i - y_i$, the partial derivative with respect to some parameter w is given by

$$\frac{\partial \hat{R}_{EXP}}{\partial w} = -2 \sum_{i=1}^{n} e_i \exp\left(\frac{e_i^2}{\tau}\right) \frac{\partial y_i}{\partial w} \ . \tag{5.35}$$

Defining as before for formulas (5.23) the weight function

$$\psi_{EXP}(e) = e \exp\left(\frac{e^2}{\tau}\right) \ , \tag{5.36}$$

we can graphically analyze the behavior of the $\hat{R}_{EXP}$ gradient. As Fig. 5.8 shows, when $\tau > 0$, ψ_{EXP} behaves in a similar way to ψ_{CE}. From small to moderate values of τ, the function has a marked hyperbolic shape: smaller errors get smaller weights with an "accelerated" trend when the errors get larger. Note again that $\lim_{\tau \to +\infty} \psi_{Exp} = \psi_{MSE}$. In conclusion, with $\hat{R}_{EXP}$ we obtain a parameterized risk functional with the flexibility to emulate a whole range of behaviors, including the ones of $\hat{R}_{ZED}$, $\hat{R}_{MSE}$ and $\hat{R}_{CE}$.

The multi-class version of $\hat{R}_{ZED}$ is given by

$$\hat{R}_{EXP} = \sum_{i=1}^{n} \tau \exp\left(\frac{e_i^T e_i}{\tau}\right) = \sum_{i=1}^{n} \tau \exp\left(\frac{1}{\tau} \sum_{k=1}^{c} e_{ik}^2\right) \ , \tag{5.37}$$

where e_{ik} is the error at the k-th output produced by the i-th input pattern. Formula (5.37) resembles for $\beta = 0$ the one proposed by Møller [160] and defined as

$$\hat{R}_{Moller} = \frac{1}{2} \sum_{i=1}^{n} \sum_{k=1}^{c} \exp\left(-\alpha(y_{ik} - t_{ik} + \beta)(t_{ik} + \beta - y_{ik})\right) \ . \tag{5.38}$$

In $\hat{R}_{Moller}$, β is the width of a region, $\mathcal{R}$, of acceptable error around the desired target and α controls the steepness of the risk outside $\mathcal{R}$. Both parameters are positive. If we increase α then $\hat{R}_{Moller}$ becomes more steep outside $\mathcal{R}$ forcing the outputs towards the boundary of that region. By decreasing β, the outputs are pulled towards the desired targets (see [160] for a detailed discussion). This risk functional was proposed in the framework of *monotonic* risk functionals [90]. In short, one can say that a risk is monotonic if its minimization (or maximization) implies the minimization of the number of misclassifications (recall the discussion in Example 2.6 where we have shown that L_{MSE} is non-monotonic). Møller defines his functional as being *soft-monotonic*, where the degree of monotonicity is controlled by α, becoming monotonic when $\alpha \to +\infty$.

We now point out the main differences between both risks. While in $\hat{R}_{Moller}$ we compute a sum (over k) of the exponentials of the squared errors, in $\hat{R}_{EXP}$ we compute the exponential of the sum (over k) of the squared errors. This brings a significant difference in terms of the gradients. In fact, for $\beta = 0$, we have

$$\frac{\partial \hat{R}_{EXP}}{\partial w} = -2 \sum_{i=1}^{n} e_{ik} \exp\left(\frac{1}{\tau} \sum_{k=1}^{c} e_{ik}^2\right) \frac{\partial y_{ik}}{\partial w}, \tag{5.39}$$

$$\frac{\partial \hat{R}_{Moller}}{\partial w} = -\sum_{i=1}^{n} \alpha e_{ik} \exp\left(\alpha e_{ik}^2\right) \frac{\partial y_{ik}}{\partial w}. \tag{5.40}$$

Thus, with $\hat{R}_{EXP}$ the backpropagated error through the output y_k uses information from *all* the other outputs, while $\hat{R}_{Moller}$ only uses the error associated to that particular output.

5.2.1.1 Gradient Descent

A gradient descent optimization can be applied to minimize $\hat{R}_{EXP}$ for any $\tau \neq 0$. Note that for $\tau < 0$ we get a negative scaled version of $\hat{R}_{ZED}$ that should now be minimized.

Algorithm 5.2 — $\hat{R}_{EXP}$ Gradient descent algorithm

1. Choose a random initial parameter vector, $\mathbf{w}$ (with components w_k).
2. Compute $\hat{R}_{EXP}$ using the classifier outputs $y_i = \varphi(\mathbf{x}_i; \mathbf{w})$ at the n available instances.
3. Compute the partial derivatives of $\hat{R}_{EXP}$ with respect to the parameters:

$$\frac{\partial \hat{R}_{EXP}}{\partial w_k} = 2 \sum_{i=1}^{n} e_i \exp\left(\frac{e_i^2}{\tau}\right) \frac{\partial e_i}{\partial w_k} \tag{5.41}$$

where $\partial e_i / \partial w_k$ depends on the classifier architecture.

4. Update at each iteration, m, the parameters $w_k^{(m)}$ using a η amount (learning rate) of the gradient:

$$w_k^{(m)} = w_k^{(m-1)} - \eta \left. \frac{\partial \hat{R}_{EXP}}{\partial w_k} \right|_{w_k^{(m-1)}} . \tag{5.42}$$

5. Go to step 2, if some stopping criterion is not met.

We thus obtain, in the same line as for $\hat{R}_{ZED}$, an $O(n)$ complexity algorithm.

In the following example we apply Algorithm 5.2 to the training of a perceptron solving a two-class problem. In Chapter 6 we describe various types of (more complex) classifiers using the $\hat{R}_{ZED}$ and $\hat{R}_{EXP}$ risks and present a more complete set of experiments and comparisons with other approaches.

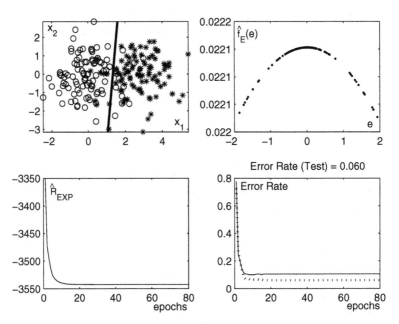

Fig. 5.9 The final converged solution of Example 5.6 with $\tau = -18$.

Example 5.6. Consider the same two-class problem of Example 5.4 (same data), now solved by a perceptron minimizing the $\hat{R}_{EXP}$ risk. We consider two values for τ: $\tau = -18$ and $\tau = 2$. Recall from formula (5.34) that minimizing $\hat{R}_{EXP}$ with $\tau = -18$ is equivalent to maximize a scaled version of $\hat{R}_{ZED}$ with $h = 3$. Figs. 5.9 and 5.10 show the final converged solution after 80 epochs. We point out the fast convergence of $\hat{R}_{EXP}$ (in about 30 epochs) mainly when we compare $\hat{R}_{ZED}$ in Fig. 5.4 with its $\hat{R}_{EXP}$ version in Fig. 5.9. Moreover, $\hat{R}_{EXP}$ reaches a good solution in terms of generalization.

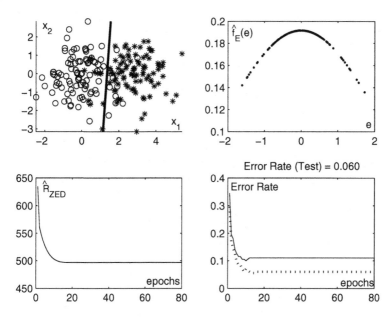

Fig. 5.10 The final converged solution of Example 5.6 with $\tau = 2$.

Chapter 6
Applications

The present chapter describes applications of error entropy (and entropy-inspired) risks, in a variety of classification tasks performed by more sophisticated machines than those considered in the preceding chapters. These include multi-layer perceptrons (MLPs), recurrent neural networks (RNNs), complex-valued neural networks (CVNNs), modular neural networks (MNNs), and decision trees. We also present a clustering algorithm using a MEE-like concept, LEGClust, which is used in building MNNs. Besides implementation issues, an extensive set of experimental results and comparisons to non-EE approaches are presented. Since the respective learning algorithms use the empirical versions of the risks, the corresponding acronyms (MSE, CE, SEE, and so forth) labeling tables and graphs of results refer from now on to the empirical versions of the risks.

6.1 MLPs with Error Entropy Risks

There are many ways to apply the entropy concept to neural networks, particularly to MLPs. These are exemplified by the works that use entropy to define and determine the complexity of a neural network [24, 57, 250], to generate a neural network [105, 226], and to perform optimization and pruning of a given neural network architecture [170, 167].

The first applications of *error entropy* as an NN risk functional to be minimized (the MEE approach), were developed by Príncipe and co-workers as pointed out in Sect. 2.3.2. The extension of MEE to classification problems handled by feed-forward MLPs was first developed by Santos et al. [198, 199, 202, 205] using Rényi's quadratic entropy. The application of Shannon's entropy for the same purpose was subsequently studied by Silva et al. [217, 212].

The present section presents several results on the application of MEE to train MLPs and on the assessment of their performance in real-world

J. Marques de Sá et al.: Minimum Error Entropy Classification, SCI 420, pp. 139–213.
springerlink.com   © Springer-Verlag Berlin Heidelberg 2013

classification tasks. Besides Shannon's and Rényi's quadratic entropies of the MLP errors we will also consider the ZED and EXP risks presented in the preceding chapter.

We will use the back-propagation algorithm for the training procedure, since it has a simple implementation which is an extension of the gradient descent procedure described in Sect. 3.1.2. Although we use back-propagation for convenience reasons, nothing in the MEE approach precludes using any other training algorithm reported in the literature (such as conjugate gradient, Levenberg-Marquardt, etc.) and usually applied with classic risk functionals such as MSE.

6.1.1　Back-Propagation of Error Entropy

The error back-propagation algorithm [192] is a popular algorithm for weight updating of MLPs when these have perceptrons (also known as neurons) with differentiable activation functions. It can be seen as an extension of the gradient descent algorithm, presented in Sect. 3.1.2, to MLP intermediary layers of perceptrons between inputs and outputs, the so-called *hidden layers*. A complete description of the back-propagation algorithm can be found in [95, 26]. In the following we only present a brief explanation of this algorithm with the purpose of elucidating its formulas for empirical entropy risks. We restrict ourselves to the one-hidden-layer case; extension to additional layers presents no special difficulties.

Figure 6.1 shows a schematic drawing of a one-hidden-layer MLP. The large circles represent perceptrons. The hidden-layer perceptrons receive input vectors with components x_{im} which are multiplied by weights and submitted to the activation function; Figure 6.1 shows the lth hidden perceptron with weights w_{ml} producing a continuous output u_{il}. This is multiplied by weight w_{lk}, and together with other hidden-layer outputs, feeds into the kth output perceptron. Each output with the known target variable generates error signals used to adjust the weights.

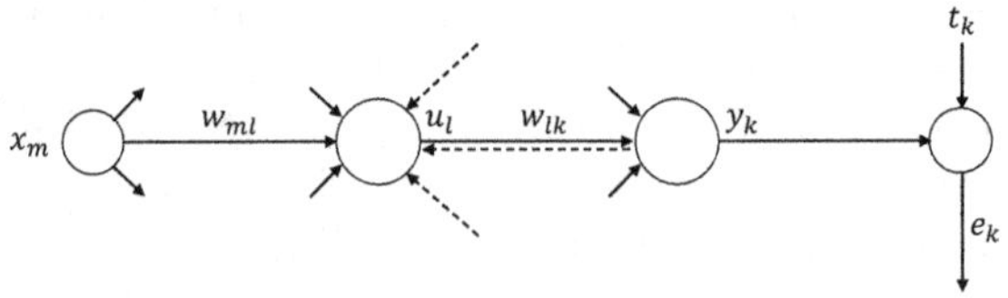

Fig. 6.1 The signal flow and the back-propagated errors (doted lines) in a one-hidden-layer MLP.

Let us start by considering the kth output perceptron with its weights being adjusted by gradient descent. When using the empirical Shannon's entropy of the error, $\hat{H}_S$, we then apply expression (3.8), which we rewrite below for the kth output perceptron in vector notation:

$$\frac{\partial \hat{H}_S}{\partial \mathbf{w}_k} = \frac{1}{n^2 h^2} \sum_{i=1}^{n} \sum_{j=1}^{n} \frac{G_h\left(\mathbf{e}_i - \mathbf{e}_j\right)}{\hat{f}(\mathbf{e}_i)} \left(e_{ik} - e_{jk}\right) \left[\frac{\partial e_{ik}}{\partial \mathbf{w}_k} - \frac{\partial e_{jk}}{\partial \mathbf{w}_k}\right]. \quad (6.1)$$

Whereas expression (3.8) contemplated the adjustment of a single weight, we now formulate the adjustment with respect to a whole vector of weights (including biases): the weight vector $\mathbf{w}_k$ of an arbitrary kth output perceptron. The derivative of $\hat{H}_S$ with respect to the weights depends on n c-dimensional error vectors denoted $\mathbf{e}_i$ and $\mathbf{e}_j$. Each component $\partial \hat{H}_S / \partial w_{lk}$ of vector $\partial \hat{H}_S / \partial \mathbf{w}_k$ in (6.1) can be conveniently expressed (namely, for implementation purposes) as the sum of all elements of the matrix resulting from:

$$\frac{1}{n^2 h^2} \begin{bmatrix} \frac{1}{\hat{f}(\mathbf{e}_1)} & \cdots & \frac{1}{\hat{f}(\mathbf{e}_1)} \\ \vdots & & \vdots \\ \frac{1}{\hat{f}(\mathbf{e}_n)} & \cdots & \frac{1}{\hat{f}(\mathbf{e}_n)} \end{bmatrix} .\times \begin{bmatrix} G_h(\mathbf{e}_1 - \mathbf{e}_1) & \cdots & G_h(\mathbf{e}_1 - \mathbf{e}_n) \\ \vdots & & \vdots \\ G_h(\mathbf{e}_n - \mathbf{e}_1) & \cdots & G_h(\mathbf{e}_n - \mathbf{e}_n) \end{bmatrix}$$

$$.\times \begin{bmatrix} e_{1k} - e_{1k} & \cdots & e_{1k} - e_{nk} \\ \vdots & & \vdots \\ e_{nk} - e_{1k} & \cdots & e_{nk} - e_{nk} \end{bmatrix} .\times \begin{bmatrix} \frac{\partial e_{1k}}{\partial w_{lk}} - \frac{\partial e_{1k}}{\partial w_{lk}} & \cdots & \frac{\partial e_{1k}}{\partial w_{lk}} - \frac{\partial e_{nk}}{\partial w_{lk}} \\ \vdots & & \vdots \\ \frac{\partial e_{nk}}{\partial w_{lk}} - \frac{\partial e_{1k}}{\partial w_{lk}} & \cdots & \frac{\partial e_{nk}}{\partial w_{lk}} - \frac{\partial e_{nk}}{\partial w_{lk}} \end{bmatrix} \quad (6.2)$$

where '$.\times$' denotes element-wise product [212]. The first matrix is not present when Rényi's quadratic entropy or information potential is used (see also expression (3.9)).

Once all n error vectors (for the n input vectors $\mathbf{x}_i$), relative to the mth training epoch have been obtained, one is then able to compute the updated weights for the output perceptron:

$$\mathbf{w}_k^{(m)} = \mathbf{w}_k^{(m-1)} - \Delta \mathbf{w}_k^{(m-1)}, \qquad \text{with} \quad \Delta \mathbf{w}_k^{(m-1)} = \eta \left. \frac{\partial \hat{H}_S}{\partial \mathbf{w}_k} \right|_{\mathbf{w}^{(m-1)}}, \quad (6.3)$$

where η is the learning rate.

The updating of the weight vector $\mathbf{w}_l$, relative to an arbitrary lth perceptron of the hidden-layer, is done as usual with the back-propagation algorithm. One needs all back-propagated errors from the output layer (incident dotted arrows in Fig. 6.1). Denoting by $\varphi(.)$ the activation function assumed the same for all perceptrons, the updating vector for $\mathbf{w}_l$ at the mth training epoch is then:

$$\Delta \mathbf{w}_l^{(m-1)} = \eta \frac{1}{n^2 h^2} \sum_{i=1}^{n} \sum_{j=1}^{n} \sum_{k=1}^{c} \frac{G_h\left(\mathbf{e}_i - \mathbf{e}_j\right)}{\hat{f}(\mathbf{e}_i)} (e_{ik} - e_{jk}) \left[\frac{\partial e_{ik}}{\partial \mathbf{w}_l} - \frac{\partial e_{jk}}{\partial \mathbf{w}_l} \right] ,$$

$$(6.4)$$

with $\frac{\partial e_{ik}}{\partial \mathbf{w}_l} = -\varphi' \left(\sum_{l=0}^{s} w_{lk} u_{il} \right) . w_{lk} \varphi' \left(\sum_{m=0}^{d} w_{ml} x_{im} \right) \mathbf{x}_i$.

Sometimes a so-called moment factor, dependent on weight differences in consecutive epochs, is added to expressions (6.3) and (6.4) with the intent to speed-up convergence (see e.g. [212]). We will not make use of the moment factor and will instead use other means to be explained later for improving convergence.

Note that the back-propagation formulas for the ZED and EXP risks are quite simpler than the above ones, essentially because there is no double sum on the errors. In fact, one has, for example

$$\frac{\partial \hat{R}_{EXP}}{\partial \mathbf{w}_k} = \sum_{i=1}^{n} e^{-\frac{1}{2\tau} \mathbf{e}_i^T \mathbf{e}_i} e_{ik} \frac{\partial e_{ik}}{\partial \mathbf{w}_k} , \qquad (6.5)$$

which gives us an equivalent complexity of these risks when compared to MSE which has

$$\frac{\partial \hat{R}_{MSE}}{\partial \mathbf{w}_k} = \frac{2}{n} \sum_{i=1}^{n} e_{ik} \frac{\partial e_{ik}}{\partial \mathbf{w}_k} . \qquad (6.6)$$

We now present an example from [198] of an MLP using Rényi's quadratic entropy trained with the back-propagation algorithm to discriminate a 4-class dataset. The example illustrates the convergence towards Dirac-δ error densities (see 3.1.1). In this example and throughout the present section we only use one-hidden-layer MLP architectures denoted $[d : n_h : c]$, with n_h the number of hidden neurons. A 1-of-c coding scheme of the outputs is assumed.

Example 6.1. Consider the two-dimensional artificial dataset shown in Fig. 6.2 consisting of 200 data instances in four separable classes with 52, 54, 42 and 52 instances in each class.

MLPs with one hidden layer, tanh activation function, and initial random weights in $[-0.1, 0.1]$, were trained using the R₂EE risk functional. Only half of the dataset (a total of 100 instances with approximately 25 instances per class) was used in the training process.

Figures 6.3, 6.4 and 6.5 show, for one experiment with $n_h = 2$, error graphs corresponding to training epochs 1, 10 and 40 respectively. Since we have a neural network with four outputs, the error vectors, e_k, $k = 1, \ldots, 4$, form a 100×4 matrix. Each figure shows a 4×4 array whose off-diagonal cells are the $(\mathbf{e}_i, \mathbf{e}_k)$ scatter plots of the column-class k error values (in $[-2, 2]$) versus the row-class i error values. The diagonal cells contain the histograms of each column-class error vector $\mathbf{e}_k$.

Analyzing these graphs one can see that the errors converge to Dirac-δ distributions and moreover with uncorrelated errors for the four classes. □

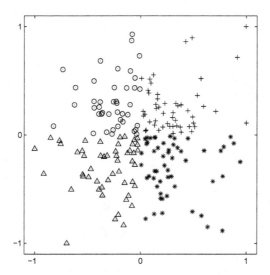

Fig. 6.2 A dataset with four separable classes.

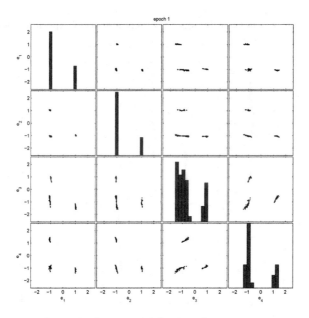

Fig. 6.3 Error graphs in the first epoch of a classification experiment with Fig. 6.2 dataset.

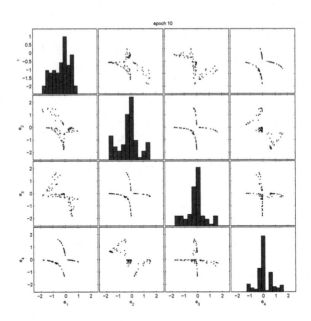

Fig. 6.4 Error graphs in the tenth epoch of a classification experiment with Fig. 6.2 dataset.

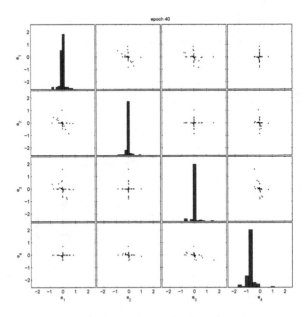

Fig. 6.5 Error graphs in the fortieth epoch of a classification experiment with Fig. 6.2 dataset.

Example 6.2. This is a continuation of Example 6.1 where several MLPs were trained and tested 40 times, using R_2EE and also MSE. The number of neurons in the hidden layer, n_h, was varied from 3 to 6. Error rate assessment was made with the holdout method: two runs with half the dataset used for training and the other half for testing, swapping the roles in the second run. The average test errors in the 40 experiments are shown in Table 6.1 where one can see that the classification test errors are always smaller with R_2EE than with MSE, being the difference highly significant [1] for $n_h = 5, 6$. □

Table 6.1 Mean test error (%) and standard deviations (in parentheses) for Example 6.2 with Student-t test probability (p). Statistically significant lower errors in bold.

n_h	R_2EE	MSE	p
3	2.43(1.33)	2.93(1.46)	0.057
4	2.20(1.20)	2.55(1.24)	0.102
5	**2.01**(1.09)	2.64(1.13)	0.007
6	**2.09**(1.02)	2.91(1.73)	0.006

In the following sub-sections we present three procedures that have in view to attain a near optimal performance and at the same time to speed up the learning process. This last issue is of particular relevance given that the algorithms using EE risk functionals are of higher complexity than those using classic risk functionals.

6.1.1.1 The Smoothing Parameter

We have seen in Chap. 3 how the choice of smoothing parameter (kernel bandwidth), h, influences the convergence towards the MEE solution. The smoothing parameter is also recognized in [198] as one of the most important factors influencing the final results of a classification problem when using MLPs trained with MEE.

As seen in Appendix E, the choice of h for optimal density estimation by the Parzen window method depends on the number of available samples and the PDF one wishes to estimate, as well as on the kernel function which we always assume to be Gaussian. Since what one needs is not an optimal PDF estimation but its fat estimation as described in Sect. 3.1.3, some latitude on the choice of h is consented; one just requires a not too large value of h above the optimum value, for the reasons pointed out in preceding chapters (see, namely, what we said in Example 3.2 and the last paragraph of Sect. 4.1.3.1).

---

[1] Statistical significance is set at the usual 5% throughout this chapter.

In the work of Santos [203] an empirically tuned formula for the choice of an h value affording a convenient fat estimation of the error PDF was proposed. The formula is inspired by another one proposed in [31] for the choice of an optimal h value in the IMSE sense[2]:

$$h_{opt} = s \left(\frac{4}{(d+2)n} \right)^{\frac{1}{d+4}} , \qquad (6.7)$$

where s is the sample standard deviation of the data and d the dimensionality of the PDF. For the estimation of the error PDF of c classes one should substitute d for c in formula (6.7). For the fat estimation of the error PDF the cited work [203] proposed, based on a large number of experiments with different datasets, the following formula for h giving higher values than (6.7) but with a similar behavior:

$$h_{fat} = 25 \sqrt{\frac{c}{n}} . \qquad (6.8)$$

Note that $nh_{fat} \to \infty$ as required (see Appendix E), and increases with c. A comparison between the values of h_{fat} and h_{opt} for different values of c is shown in Fig. 6.6. Note that $h_{fat} > h_{opt}$ specially for small values of n.

Table 6.2 summarizes results of experiments reported in the cited work [203], where details on the datasets are provided. The "Best h" column presents the values of h achieving the minimum number of classification errors in 20 experiments; the "Suggested h" column presents the averages of the h values achieving the 10 smallest classification errors. For comparison, Table 6.2 also shows the values supplied by formulas (6.7) and (6.8).

Table 6.2 Values of h for the experiments reported in [203] and those obtained with formulas (6.7) and (6.8).

Datasets	c	n	Best h	Suggested h	h_{fat}	h_{opt}
Ionosphere	2	351	4.6	3.9	2.67	0.85
Sonar	2	208	3.4	3.7	3.47	0.92
Wdbc	2	569	1.4	1.7	2.10	0.78
XOR-n	2	200	4.0	4.7	3.54	0.93
Iris	3	150	4.0	3.6	5.00	1.05
Wine	3	178	4.6	4.2	4.59	1.02
PB12	4	608	1.8	2.2	2.87	0.93
XOR-n	4	200	5.2	5.0	5.00	1.07
Olive	9	572	4.6	5.4	4.43	1.20

---

[2] Note that for the univariate case ($d = 1$) formula (6.7) reproduces formula (E.19), when using the sample estimate of the standard deviation.

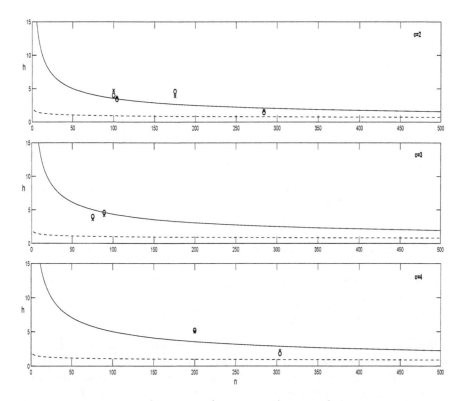

Fig. 6.6 Value of h_{opt} (dashed line) and h_{fat} (solid line) for $c = 2$, 3, and 4. (Marked instances refer to experiments with datasets summarized in Table 6.2; see text.)

Figure 6.6 shows the h values obtained in these experiments (Table 6.2), with open circles for the "Best h" and crosses for the "Suggested h". The h_{fat} values proposed by formula (6.8) closely follow these values.

One could think of making h an updated variable along the training process, namely proportional to the error variance. This is suggested by the fact that, as one approaches the optimal solution, the errors tend to concentrate making sense to decrease h for the corresponding smaller data range. However, it turns out to be a tricky task to devise an empirical updating formula which does not lead to instability in the vicinity of the minimum error, given the wild variations that a small h value originates in the gradient calculations. Figure 6.7 shows an MLP training error curve for the PB12 dataset [110] using an updated h, which is simply proportional to the error variance. Also shown is the value of h itself. Near 100 epochs and with training error 10% the algorithm becomes unstable loosing the capability to converge.

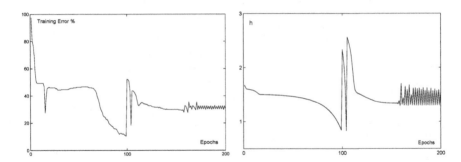

Fig. 6.7 Instability of the MEE back-propagation algorithm, when updating h in an experiment with PB12 dataset.

6.1.1.2   Adaptive Learning Rate

Several authors, following different strategies, have shown that, by adapting the value of the learning rate, η, along the learning process, one can get a better and faster convergence of a neural network using the MSE risk functional [20, 232, 36, 109, 186, 210, 72, 104, 145, 146].

We now show how η can be adjusted as a function of the error entropy, in a similar way as when using MSE, using the following simple but effective rule: if the error entropy decreases between two consecutive epochs of the training process, the algorithm then produces an increase in the learning rate parameter; on the other hand, if the error entropy increases between two consecutive epochs, then the algorithm produces a decrease in the learning rate parameter and, furthermore, the updating step is restarted, i.e., we recover the previous "good" values of all neural network weights. This simple rule for learning rate updating can be written as [205]:

$$
\eta^{(m)} = \begin{cases} \eta^{(m-1)}u & \text{if } \ H^{(m)} < H^{(m-1)} \\ \eta^{(m-1)}d \ \wedge \text{restart} & \text{if } \ H^{(m)} \geq H^{(m-1)} \end{cases}, \quad u > 1, \ 0 < d < 1 ,
$$

$$(6.9)$$

where $\eta^{(m)}$ and $H^{(m)}$ are, respectively, the learning rate and the error entropy at the mth iteration and u and d are the increasing and decreasing updating factors.

A large number of experiments, reported in [205], were performed in order to find adequate values for u and d; based on the respective results discussed in detail in [205] the following values for u and d were proposed: $u = 1.2$ and $d = 0.2$. Figure 6.8 shows the results of one such experiment with a real-world dataset, namely the evolution of the training error and of the error entropy using both fixed learning rate (FLR) and variable learning rate (VLR) rule (6.9) using the proposed values of u and d. Clearly, the smallest classification error

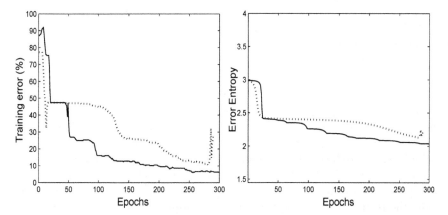

Fig. 6.8 Training Error (left) and Error Entropy (right) curves for FLR (dotted line) and VLR (solid line) in the classification of a real-world dataset [205].

was attained when using VLR and moreover with a continuous decrease of the error entropy.

6.1.2   The Batch-Sequential Algorithm

Error entropy estimation based on the Parzen window method implies using all available error samples. The use of the batch mode in the back-propagation algorithm is therefore obligatory. It is known, however, that the batch mode — weight updating after the presentation of all training set samples — has some limitations over the sequential mode — weight updating after the presentation of each training set sample [95]. To overcome these limitations a brief reference to the possibility of combining both batch and sequential modes when training neural networks was made in [26]. An algorithm combining the two modes, trying to capitalize on their mutual advantages, was effectively proposed in [202] and tested within the specific MEE scope with Rényi's quadratic entropy: MEE Batch-Sequential algorithm, MEE-BS.

One of the advantages of the batch mode is that the gradient vector is estimated with more accuracy, guaranteeing the convergence to, at least, a local minimum. The sequential mode of weight updating leads to a sample-by-sample *stochastic* search in the weight space implying that it becomes less likely for the back-propagation algorithm to get trapped in local minima [95]. However, for EE risks one still needs a certain quantity of data samples to estimate the entropy, and this limits the use of the sequential mode.

One way to overcome this dilemma, proposed in [202], consists of splitting the training set into several groups that are presented to the algorithm in a sequential way. The batch mode is applied to each group.

Let $U = (X_{ds}, T_{ds})$ denote the training set extracted from a given dataset and U_j the subsets obtained by randomly dividing U into several groups with an equal number of samples, such that

$$|U| = r + \sum_{j=1}^{L} |U_j| \, , \qquad (6.10)$$

where L is the number of subsets and r the remainder. This division is performed in each epoch of the learning phase. The partition of the training set into subsets reduces the probability of the algorithm getting trapped in local minima since it is performed in a random way. The subsets are sequentially presented to the learning algorithm, which applies to each one, in batch mode, the respective back-propagation and subsequent weight update.

One of the advantages of using the batch-sequential algorithm is the decrease of algorithm complexity. The complexity of the original MEE algorithm is $O(|U|^2)$; with the MEE-BS algorithm the complexity is proportional to $L(|U|/L)^2$, which means that, in terms of computational time, one achieves a reduction proportional to L. The number of subsets, L, is influenced by the size of the training set. One should avoid using subsets with a number of samples less than 40 [202].

Classifier performance using the MEE-BS algorithm seems to be quite insensitive to "reasonable" choices of the number of subsets, L. Table 6.3 shows the best results in experiments reported in [202] on four real-world datasets. No statistically significant variation of the error rate statistics (mean and standard deviation in 20 repetitions of the hold-out method) were found for two different values of L. The same conclusion was found to hold when the number of epochs and of hidden neurons used in these experiments were varied. In what concerns the processing time per epoch, as also shown in Table 6.3, the MEE-BS algorithm was found to be up to six times faster than the MEE-VLR algorithm.

The batch-sequential algorithm can also be implemented with variable learning rate. However, the simple "global" updating rule described in the previous section cannot be applied. The reason for this is easy: since in MEE-VLR we compare the error entropy of a certain epoch with its value in the previous one for the *same* samples, we cannot apply it to the batch-sequential algorithm because, in each epoch, we use *different* sets.

Instead of using the simple procedure described in the preceding section we may employ the variation of the learning rate in such a way that it is done by comparing the respective *gradient* in consecutive iterations. Two learning rate updating rules can then be incorporated into the MEE-BS algorithm: either the Silva and Almeida's rule [210] (MEE-BS(SA)) or the resilient back-propagation rule [186] (MEE-BS(RBP)). Both variants of the MEE-BS algorithm are described in detail in [202]. An example of the training phase using the three methods (MEE-BS, MEE-BS(SA), and MEE-BS(RBP)) is shown in Fig. 6.9.

Table 6.3 Best results of MEE-BS and MEE-VLR algorithms with the time per epoch, T, in 10^{-3} sec.

	Error (Std)	L	n_h	Epochs	T	T_{VLR}/T_{BS}
Dataset "Ionosphere" ($n=351$, $c=2$)						
	Error (Std)	*L*	*n_h*	*Epochs*	*T*	*T_{VLR}/T_{BS}*
MEE-VLR	12.06(1.11)	-	12	40	16.7	-
MEE-BS	12.27(1.23)	4	8	100	6.4	2.6
MEE-BS	12.00(1.09)	8	8	100	4.8	3.5
Dataset "Olive" ($n=572$, $c=9$)						
	Error (Std)	*L*	*n_h*	*Epochs*	*T*	*T_{VLR}/T_{BS}*
MEE-VLR	5.04(0.53)	-	25	200	77.7	-
MEE-BS	5.17(0.51)	4	30	140	17.6	4.4
MEE-BS	5.24(0.70)	8	20	180	12.8	6.1
Dataset "Wdbc" ($n=569$, $c=2$)						
	Error (Std)	*L*	*n_h*	*Epochs*	*T*	*T_{VLR}/T_{BS}*
MEE-VLR	2.33(0.37)	-	4	40	38.7	-
MEE-BS	2.31(0.35)	5	10	60	13.6	2.8
MEE-BS	2.35(0.48)	8	10	40	9.6	4.0
Dataset "Wine" ($n=178$, $c=3$)						
	Error (Std)	*L*	*n_h*	*Epochs*	*T*	*T_{VLR}/T_{BS}*
MEE-VLR	1.83(0.83)	-	14	40	5.8	-
MEE-BS	1.88(0.80)	4	16	60	3.2	1.8
MEE-BS	1.88(0.86)	8	16	60	2.5	2.3

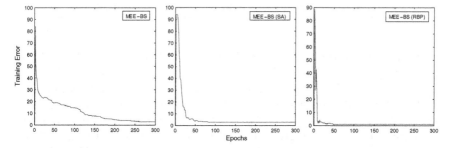

Fig. 6.9 Training curves for MEE-BS, MEE-BS(SA) and MEE-BS(RBP) algorithms in one experiment with a real-world dataset.

6.1.3 Experiments with Real-World Datasets

Experiments comparing MLPs with different risk functionals, applied to the classification of real-world datasets, were reported in several works.

In the experiments reported in [198], datasets Diabetes, Wine and Iris (UCI repository [13]) were used. Several MLPs with varying n_h were trained and tested with the holdout method using H_{R_2} and MSE. The results of these experiments are presented in Table 6.4. The superiorness of R$_2$EE over MSE for these datasets is clear. The bottom row of Table 6.4 also hints a stabler design of R$_2$EE than MSE.

Table 6.4 Mean (standard deviation) of test error (%) for the experiments reported in [198]. Last row is the standard deviation of the mean errors for all n_h values. Best results in bold.

n_h	Diabetes		Wine		Iris	
	R$_2$EE	MSE	R$_2$EE	MSE	R$_2$EE	MSE
2	23.80(0.94)	28.40(4.87)	3.62(1.30)	9.72(10.60)		
3	23.94(0.97)	27.25(4.72)	3.81(1.00)	4.27(3.77)	4.36(1.12)	4.72(4.75)
4	23.99(1.52)	26.42(4.53)	**1.94**(0.72)	3.03(1.08)	4.43(1.30)	4.75(1.27)
5	23.80(1.04)	25.10(1.80)	2.50(1.01)	3.20(1.83)	4.38(1.34)	4.15(1.32)
6	24.10(1.33)	24.70(1.80)	2.47(1.20)	3.06(1.43)	4.30(1.16)	**3.97**(1.05)
7	24.10(0.90)	24.40(1.06)	2.44(1.00)	2.39(1.50)	4.41(1.42)	5.18(4.74)
8	23.90(0.71)	23.90(1.18)	2.16(0.92)	2.92(1.07)	4.31(1.27)	4.65(1.32)
9	24.30(1.42)	24.00(0.95)	2.22(0.83)	2.50(1.35)		
10	**23.60**(0.86)	24.10(1.20)	2.31(0.51)	2.95(1.29)		
11	24.02(1.00)	27.41(5.19)				
12	24.93(3.24)	27.64(5.04)				
STD	**0.35**	1.69	**0.65**	2.29	**0.05**	0.44

Experiments on two-class artificial and real-world datasets with MLPs using EE risks (R$_2$EE, SEE), EXP, and the classical risks MSE and CE, were reported in [215]. The datasets are shown in Table 6.5. Six of them are artificial checkerboard datasets, an example of which is shown in Fig. 6.10; both 2×2 (CB2x2) and 4×4 (CB4x4) "boards" are used, with some percentage of the less represented class given in parenthesis (e.g., CB2x2(10) has 10% of the total number of instances assigned to one of the classes). The other six datasets are real-world and available in [13].

All experiments with a given dataset used the same MLP architecture with n_h guaranteeing acceptable generalization. Regularization was performed by early stopping (see, e.g., [26] on this issue) using a preliminary suite of 10 runs in order to choose an adequate number of epochs as well as an adequate value of h (starting with initial values given by formula (6.8)). Details on all these issues are given in [215].

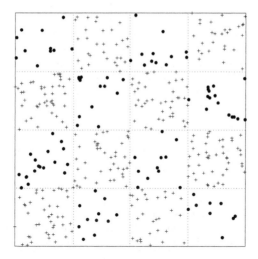

Fig. 6.10 A 4×4 checkerboard dataset with 400 instances (100 instances in the minority class: dots). Dotted lines are for visualization purpose only.

Table 6.5 The datasets of the experiments reported in [215].

	CB2×2(50)	CB2×2(25)	CB2×2(10)	CB4×4(50)	CB4×4(25)	CB4×4(10)
No. cases	200	400	1000	200	400	1000
No. feat.	2	2	2	2	2	2
	Cl. Hea. D.	Diabetes	Ionosphere	Liver	Sonar	Wdbc
No. cases	207	768	351	345	208	569
No. feat.	13	8	34	6	60	30

The performance assessment was based on 20 runs of the holdout method and computation of the following statistics (all in percentage) in each run:

Correct classification rate: $P_{ct} = 1 - P_{et}$, with P_{et} defined as in Sect. 3.2.2.

Balanced correct classification rate, P_{bct}, a variant of P_{ct} computed with the equal prior assumption.

Area under the Receiver Operating Characteristic curve, which measures the trade-off between sensitivity and specificity in two-class decision tables (for details, see e.g. [149]). The higher the area the better the decision rule.

The last two performance measures are especially suitable for unbalanced datasets, such as the artificial checkerboard datasets, where an optimistically high P_{ct} could arise from a too high sensitivity or specificity. P_{ct} was only used for the real-world datasets.

The mean values of P_{ct} and P_{bct} (in the 20 runs) with the standard deviations are shown in Table 6.6. The highest mean value (a "win") appears in bold. The Friedman test and the multiple sign test, adequate for multi-comparison assessment of classifiers [51, 79, 102, 195], found no statistically significant difference among the five risk functionals at the usual 5% significance level ($p = 0.1$ for the Friedman test). The area under the Receiver Operating Characteristic curve didn't alter these conclusions for the checkerboard datasets. The counts of wins suggest a superior performance of CE and EXP; the number of datasets is, however, too small to afford more definite conclusions.

Table 6.6 Percent mean (standard deviation) values of $\bar{P}_{bct}$ and $\bar{P}_{ct}$ with 'wins' (bold).

	$\bar{P}_{bct}$					
	CB2×2(50)	CB2×2(25)	CB2×2(10)	CB4×4(50)	CB4×4(25)	CB4×4(10)
MSE	92.87 (2.48)	92.47 (4.73)	76.97 (6.51)	77.94 (4.39)	76.56 (3.96)	70.77 (3.63)
CE	91.25 (2.50)	92.87 (2.35)	83.40 (3.69)	**79.40** (3.20)	**82.98** (1.78)	80.49 (1.98)
EXP	**92.94** (4.30)	93.29 (3.61)	**94.14** (5.16)	78.54 (3.57)	80.21 (3.05)	**81.47** (3.64)
SEE	86.96 (3.17)	**94.36** (1.80)	90.22 (5.66)	73.58 (4.96)	70.93 (4.03)	67.98 (3.95)
R$_2$EE	92.48 (3.17)	90.70 (5.45)	91.80 (4.87)	75.59 (5.45)	71.20 (3.69)	73.04 (3.71)

	$\bar{P}_{ct}$					
	Clev. H. D.	Diabetes	Ionosphere	Liver	Sonar	Wdbc
MSE	82.42 (1.08)	76.58 (0.88)	87.81 (1.21)	68.52 (1.97)	78.82 (2.51)	97.39 (0.67)
CE	**83.33** (1.07)	**76.82** (0.77)	88.04 (1.46)	69.04 (2.01)	78.70 (2.60)	**97.44** (0.55)
EXP	81.72 (1.29)	76.76 (0.79)	88.37 (1.88)	69.80 (1.27)	78.82 (2.92)	97.21 (0.68)
SEE	82.72 (1.11)	76.66 (0.85)	87.71 (1.37)	69.08 (1.96)	**79.18** (2.50)	97.36 (0.66)
R$_2$EE	81.77 (1.81)	75.84 (0.78)	**88.50** (1.33)	**70.32** (1.54)	77.43 (3.01)	96.89 (0.42)

Further experiments with the datasets of Table 6.5 (not described in [215]) were performed aimed at the comparison of activation functions tanh (used by default in all experiments described in this chapter), atan, and logistic sigmoid (see Sect. 3.3). The risk functional used was R$_2$EE and the experiments were run following the same protocol as in [215]. Table 6.7 shows the respective results ("sig" means "logistic sigmoid"). The logistic sigmoid was the activation function that led to poorer performances in these twelve datasets. The differences between the tanh and atan activation functions seem more moderate.

Table 6.7 Percent mean (standard deviation) values of $\bar{P}_{bct}$ and $\bar{P}_{ct}$ with 'wins' (bold).

	$\bar{P}_{bct}$					
a.f.	CB2x2(50)	CB2x2(25)	CB2x2(10)	CB4x4(50)	CB4x4(25)	CB4x4(10)
tanh	**92.48** (3.17)	**90.70** (5.45)	**91.80** (4.87)	**75.59** (5.45)	71.20 (3.69)	**73.04** (3.71)
sig	86.16 (4.78)	85.03 (10.93)	68.08 (6.58)	72.21 (4.83)	70.87 (6.76)	68.79 (6.86)
atan	86.40 (6.81)	90.25 (4.02)	83.71 (8.25)	70.85 (6.54)	**72.89** (6.79)	60.63 (6.08)

	$\bar{P}_{ct}$					
a.f.	Clev. H. D.	Diabetes	Ionosphere	Liver	Sonar20	Wdbc
tanh	81.77 (1.81)	**75.84** (0.78)	**88.50** (1.33)	**70.32** (1.54)	77.43 (3.01)	**96.89** (0.42)
sig	82.25 (1.33)	73.74 (1.43)	86.06 (1.53)	67.19 (3.41)	77.60 (2.76)	96.79 (0.82)
atan	**82.87** (0.94)	75.72 (1.09)	87.70 (1.14)	68.66 (1.91)	**78.52** (3.04)	96.86 (0.61)

A more comprehensive comparison study was reported in [220], involving six risk functionals — MSE, CE, EXP, ZED, SEE, R_2EE — and 35 real-world datasets. The MLPs had the same architecture and their classification tasks were performed according to the same protocol. Twenty repetitions of the classification experiments using stratified 10-fold cross-validation were carried out for datasets with more than 50 instances per class; otherwise, 2-fold cross-validation was used. Pooled means of training set and test set errors and of their balanced counterparts — $\bar{P}_e = (\bar{P}_{ed} + \bar{P}_{et})/2$, $\bar{P}_b = (\bar{P}_{bd} + \bar{P}_{bt})/2$ — were computed, as well as the pooled standard deviations — $sP_e = (sP_{ed}^2/2 + sP_{et}^2/2)^{1/2}$, $sP_b = (sP_{bd}^2/2 + sP_{bt}^2/2)^{1/2}$.

The generalization ability was assessed in the same way as in Sect. 3.2.2, using $D_e = \bar{P}_{et} - \bar{P}_{ed}$, and $D_b = \bar{P}_{bt} - \bar{P}_{bd}$ for the balanced error counterpart.

Large tables of performance statistics and of multiple sign tests are provided in [220]. The statistical tests showed that the ubiquitous MSE was the less interesting risk functional to be used by MLPs: MSE never achieved a significantly better classification performance than competing risks. CE and EXP were the risks found by the several tests (Friedman, multiple sign, chi-square goodness-of-fit for counts of wins and losses, Wilcoxon paired rank-sum) to be significantly better than their competitors. Counts of significantly better and worse risks have also evidenced the usefulness of SEE and R_2EE for some datasets. It was namely found, in this study, that for some datasets SEE and R_2EE reached a significantly higher performance than any other risk functional; even though performance-wise they positioned between MSE and {CE, EXP}, they were "irreplaceable" for some datasets. This was not evidenced by other risk functionals: the highest performing risk had a comparable competitor (no statistically significant difference).

In what regards the generalization issue, it was found that all risks except R_2EE behaved similarly. R_2EE exhibited significantly poor generalization, as shown in the Dunn-Sidak [56] diagram for D_e scores of Fig. 6.11.

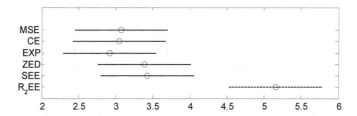

Fig. 6.11 Dunn-Sidak comparison intervals for the D_e scores.

An interesting issue regarding MLPs is their comparison with SVMs, a type of classifier characterized by optimal generalization ability, given the inherent constraint on the norm of the weight vector. Collobert and Bengio [43] elucidated the links between SVMs and MLPs; they showed that under simple conditions a perceptron is equivalent to an SVM, and moreover that the early stopping rule used in stochastic gradient descent training of MLPs is a regularization method that constraints the norm of the weight vector, and therefore its generalization ability.

Experimental evidence on the comparison of MLPs and SVMs is provided in [7]. The same datasets and methodology of [220] was followed. The MLPs used the MSE, CE, EXP and SEE risk functionals. The SVMs were implemented with a radial basis function (RBF) kernel [46].

The same battery of statistical tests of [220] was applied to the experimental results showing no significant difference among the classifiers in terms of unbalanced error rates. In terms of balanced error rates SVM-RBF performed *significantly worse* than MLP-CE and MLP-EXP. Regarding generalization, SVM-RBF and MLP-EXP scored as the classification methods with significantly better generalization, both in terms of balanced and unbalanced error rates. Thus, even in terms of generalization SVMs had worthy MLP competitors (in the studied 35 datasets).

6.2 Recurrent Neural Networks

Recurrent neural networks (RNNs), as opposed to feed-forward implementations, allow information to pass from a layer into itself or into a previous layer. This recurrent behavior implies a feedback that makes feed-forward learning algorithms, such as back-propagation, unfit for these networks. Back-propagation is generalized for RNNs in the form of back-propagation through time (BPTT) [240]. Another important learning method for RNNs is real time recurrent learning (RTRL) [241] which is discussed in a following section.

The recurrent nature raises the issue of stability [96] which manifests itself during training: not all networks with identical topology are able to learn. Depending on the initialization of their weights, some might not converge during training.

The main use of RNNs is for time dependent tasks, such as learning symbolic sequences or making time series predictions although they can also be used as associative memories.

The following two sections focus on the application of MEE to two types of RNNs, showing empirically that it can improve stability and, in some cases, the overall performance when compared to MMSE-based algorithms.

6.2.1 Real Time Recurrent Learning

6.2.1.1 Introduction

In this section the idea of the ZED risk, discussed in Chap. 5, is adapted
to learning in recurrent neural networks. It is shown how an online learning
algorithm for RNN, the Real Time Recurrent Learning (RTRL) [241] can
make use of the ZED principle. In Sect. 6.2.1.4 two applications are presented:
a symbolic prediction problem and a time series forecast problem (Mackey-
Glass chaotic time series). In both cases the new approach has advantages
when compared to the original RTRL.

6.2.1.2 RTRL

The RTRL algorithm for training fully recurrent neural networks was originally
proposed in [241]. Many modifications to this original proposal have been presented. This section follows with minor changes the notation of [95]. Consider
the fully recurrent neural network of Fig. 6.12. It contains q neurons, d inputs
and p outputs. The state-space description is given by the following equations

$$\mathbf{x}(\tau + 1) = \left[\varphi(\mathbf{w}_1^T \boldsymbol{\xi}(\tau)), \dots, \varphi(\mathbf{w}_q^T \boldsymbol{\xi}(\tau))\right]^T, \tag{6.11}$$

where $\mathbf{x}$ represents here the state vector, τ stands for the time, and φ is the
activation function. The $(q + d + 1)$ vector $\mathbf{w}_j$ contains the weights of neuron
j; and $\boldsymbol{\xi}(\tau)$ is another $(q + d + 1)$ vector defined by $[\mathbf{x}(\tau), \mathbf{u}(\tau)]^T$. The $(d + 1)$
input vector $\mathbf{u}(\tau)$ contains 1 in the first component (the bias fixed input
value); the remaining components are the d network inputs. The equation
that gives the p vector of network outputs, $\mathbf{y}$, is

$$\mathbf{y}(\tau) = \mathbf{C}\mathbf{x}(\tau), \tag{6.12}$$

where $\mathbf{C}$ is a $p \times q$ matrix that is used to select which neurons produce the
network output. The idea is to use the instantaneous gradient of the error
to guide the search for the optimal weights that minimize this error. The
algorithm works by computing the following for each time τ:

$$\boldsymbol{\Lambda}_j(\tau + 1) = \boldsymbol{D}(\tau)\left(\mathbf{W}_a(\tau)\boldsymbol{\Lambda}_j(\tau) + \mathbf{U}_j(\tau)\right), \tag{6.13}$$

$$\mathbf{e}(\tau) = \mathbf{t}(\tau) - \mathbf{C}\mathbf{x}(\tau), \tag{6.14}$$

$$\Delta\mathbf{w}_j = \eta \left(\mathbf{e}(\tau)^T \mathbf{C}\boldsymbol{\Lambda}_j(\tau)\right)^T, \tag{6.15}$$

where $\boldsymbol{\Lambda}_j$ contains the partial derivatives of $\mathbf{x}$ w.r.t. the weight vector $\mathbf{w}_j$,
$\boldsymbol{D}$ is a diagonal matrix with the partial derivatives of the activation function
w.r.t. its arguments, $\mathbf{W}_a$ contains part of the network weights and $\mathbf{U}_j$ is a
zero matrix with the transpose of vector $\boldsymbol{\xi}$ in its jth row (see [95] for details).
Vector $\mathbf{e}$ is the error and $\mathbf{t}$ the desired target output.

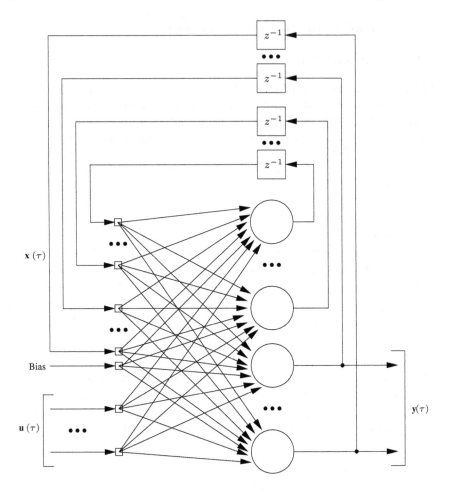

Fig. 6.12 A fully recurrent one hidden layer neural network. The notation is explained in the text.

6.2.1.3   Application of ZED to RTRL

The application of the ZED in this case cannot be based, as previously done, using a static set of n error values. Instead, one has to consider L previous values of the error in order to build a dynamic approximation of the error density. This will use a time sliding window that will allow the definition of the density in an online formulation of the learning problem.

For a training set of size n, the error r.v. $\mathbf{e}(i)$ represents the difference between the desired output vector and the actual output, for a given pattern i. The generalization of expression (5.9) to a p-dimensional output (the dimension of $\mathbf{e}$ or the number of network outputs), is given by

$$\hat{f}(\mathbf{0}; \mathbf{w}) = \frac{1}{nh^p} \sum_{i=1}^{n} K\left(\frac{\mathbf{0} - \mathbf{e}(i)}{h}\right) , \tag{6.16}$$

where h represents the bandwidth of the kernel K. Using the Gaussian kernel with zero mean and unit covariance in this expression, the estimator for the error density becomes

$$\hat{f}(\mathbf{0}; \mathbf{w}) = \frac{1}{\sqrt{2\pi}nh^p} \sum_{i=1}^{n} \exp\left(-\frac{\mathbf{e}(i)^T\mathbf{e}(i)}{2h^2}\right) . \tag{6.17}$$

Due to reasons discussed in [214] and Chap. 5, related to the speed of convergence of ZED, instead of using expression (6.17) the following simplification shall be used (in line with the EXP risk):

$$\hat{f}(\mathbf{0}; \mathbf{w}) = h^2 \sum_{i=1}^{n} \exp\left(-\frac{\mathbf{e}(i)^T\mathbf{e}(i)}{2h^2}\right) ; \tag{6.18}$$

given that the difference relies only on constant terms, the same extrema are found. The gradient of (6.18) is:

$$\frac{\partial \hat{f}(\mathbf{0}; \mathbf{w})}{\partial \mathbf{w}} = -\sum_{i=1}^{n} \exp\left(-\frac{\mathbf{e}(i)^T\mathbf{e}(i)}{2h^2}\right) \mathbf{e}(i)\frac{\partial \mathbf{e}(i)}{\partial \mathbf{w}} . \tag{6.19}$$

Since the search is for the weights yielding the maximum of the error density at $\mathbf{0}$, the network weight update shall be made by

$$\Delta \mathbf{w} = \eta\frac{\partial \hat{f}(\mathbf{0}; \mathbf{w})}{\partial \mathbf{w}} , \tag{6.20}$$

where η stands for the learning rate.

This adaptation of ZED will change expression (6.19) to

$$\frac{\partial \hat{f}(\mathbf{0}, t; \mathbf{w})}{\partial \mathbf{w}} = -\sum_{i=1}^{L} \exp\left(-\frac{\mathbf{e}(t-i)^T\mathbf{e}(t-i)}{2h^2}\right) \mathbf{e}(t-i)\left(\frac{\partial \mathbf{e}(t-i)}{\partial \mathbf{w}}\right) , \tag{6.21}$$

where instead of computing the density over the n data instances of a training set, it is computed over the last L errors of the RNN. Notice that the time dependency of the gradient of the density is now explicit. This approach is an approximation to the real gradient of the density since it uses error values from different time steps to create an estimate of the error density. Given that learning is online and the weights are adjusted at each time step, the construction of a density from errors at different time steps is valid if L is not too large, since it would then include information from a very different weight state; L can't be too small either otherwise the density estimation will suffer from a lack of samples. The modification of the RTRL algorithm is on equation (6.15) containing the weight update rule, which is rewritten as

$$\Delta \mathbf{w}_j = \eta \sum_{i=1}^{L} \exp\left(-\frac{\mathbf{e}(t-i)^T \mathbf{e}(t-i)}{2h^2}\right) \left(\mathbf{e}(t-i)^T \mathbf{C} \mathbf{\Lambda}_j(t-i)\right)^T . \quad (6.22)$$

The minus sign on expression (6.21) cancels with the minus sign that would have to be inserted into the $\Delta \mathbf{w}_j$ given that maximization is now performed instead of minimization.

6.2.1.4 Experiments

This section describes experiments conducted to evaluate the performance of the RTRL-ZED method. The obtained experimental results are compared against those of the original RTRL method. The activation function used in the RNNs is the standard sigmoid. The variables $\mathbf{x}$ and $\mathbf{\Lambda}_j$ were initialized as follows: $\mathbf{x}(0) = \mathbf{0}, \mathbf{\Lambda}_j(0) = \mathbf{0}, \ j = 1, \ldots, q$.

The first experiment uses 3000 instances of the Mackey-Glass time series [143]. The first 2000 are ignored in terms of prediction evaluation (it is considered that the network is still adapting to the signal) and the last 1000 are used for evaluation. The task consists on predicting the next point of the series with an absolute error smaller than 0.05 (the tolerance). This can be seen as a classification task if we consider that for each point that the network tries to predict, it is in fact classifying the input as a member of the Mackey-Glass series or not depending on whether it is able to predict within the prescribed tolerance. We count the number of misclassified points (predicted with absolute error higher than the tolerance) on the last 1000 points of the series.

Table 6.8 Average error in percentage (with standard deviation) for 100 repetitions of the experiment using the Mackey-Glass time series.

q	2	4	6
RTRL	4.37 (0.15)	4.53 (0.32)	4.56 (0.39)
RTRL-ZED	1.53 (0.09)	1.53 (0.05)	1.41 (0.09)

Table 6.8 presents the results. It contains the average errors and standard deviations for 100 repetitions of the experiment. The parameters were varied in the following way: for RTRL, η was varied from 1 to 20 with 0.2 steps; for RTRL-ZED, η was varied from 0.1 to 1 with 0.05 steps, h varied from

0.5 to 10 with 0.5 steps, and the number of previous values $L \in \{8, 10, 12\}$ was used. Table 6.8 only contains the best results for each network. Both networks were evaluated using 2, 4 and 6 neurons.

We can see that while the average error value was around 4.5% for the RTRL algorithm, the RTRL-ZED had errors around 1.5%. All results obtained with RTRL-ZED are statistically significantly better (smaller error) than those obtained with RTRL (t-test $p \approx 0$). It appears that only 2 neurons are enough in both cases to learn the problem, but the RTRL-ZED seams to benefit from an increase in number of neurons since the result for $q = 6$ had an error of 1.4% which is smaller than the errors obtained with less neurons.

The second experiment is adapted from [2], and consists in predicting the next symbol of the sequence: 0 1 0 0 1 0 0 0 1 0 0 0 0 1 ..., up to twenty zeros, always followed by a one. The number of symbols the network needed to see in order to correctly make the remaining predictions until the end of the sequence, was recorded. The sequence is composed of 230 symbols. A hundred repetitions were made starting with random initialization of the weights, with the learning rate, η, ranging from 5 to 39 for the standard RTRL and from 2 to 9 for RTRL-ZED; the kernel bandwidth, h, varied in $\{1, 2, 3\}$ and the size of the sliding window for the temporal estimation of the density, L, in $\{8, 10, 12\}$.

The results are shown in Fig. 6.13. Each point in these figures represents the percentage of convergence in 100 experiments versus the correspondent average number of symbols (NS) necessary to learn the problem, for the standard RTRL (star) and the RTRL-ZED (square) networks. The various points were obtained by changing the parameters η, L, and h (in the case of standard RTRL only η is used). Only the cases where at least one of the 100 repetitions converged were plotted.

The figures show that standard RTRL is not able to obtain more than 40% convergence, but the RTRL-ZED can reach 100% convergence. It can also be observed that, in general, for a given value of NS, the RTRL-ZED is able to obtain higher percentages of convergence than the original RTRL. A slight advantage of the original RTRL over the new proposal is that it is able to learn the problem with fewer symbols but by a small difference.

6.2.2 Long Short-Term Memory

Typical RNN implementations suffer from the problem of loosing error information pertaining to long time lapses. This occurs because the error signals tend to vanish over time [95]. One of the most promising machines for sequence learning, addressing this information loss issue, is the Long Short-Term Memory (LSTM) recurrent neural network [103, 82, 81, 171]. In fact, it has been shown that LSTM outperforms traditional RNNs such as Elman [63], Back-Propagation Through Time (BPTT) [242] and Real-Time

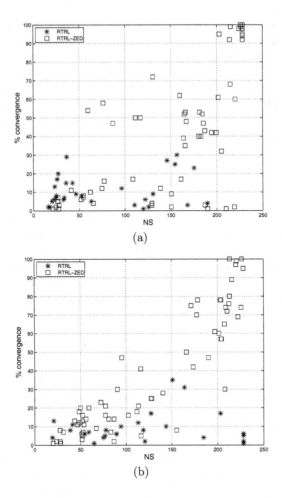

Fig. 6.13 Percentage of convergence on 100 experiments versus the correspondent average number of symbols necessary for learning the problem (NS), for the standard RTRL and the RTRL-ZED. First experiment. a) Network with 4 neurons. b) Network with 6 neurons.

Recurrent Learning (RTRL) [187] in problems where the need to retain information for long time intervals exists [103]. LSTM is able to deal with this problem since it protects error information from decaying using gates.

In this section it is shown how to adapt the LSTM for learning long time lapse classification problems using MEE and several experiments are presented showing the benefits that can be obtained from this approach.

6.2.2.1 Introduction

The LSTM network was proposed in [103]. It uses gates preventing the neuron's non-linearities (the transfer functions) from making the error information pertaining to long time lapses vanish. Note that the use of gates is only one possibility to avoid this problem that affects traditional RNNs; other approaches can be found in [95]. In the following brief discussion we consider the original LTSM approach [103]; other LSTM versions can be found in [82, 81, 171].

An overview of the LSTM network is in Fig. 6.14: it is a RNN with the hidden layer feeding back into itself as well as into the output layer. Apart from the usual neurons, the hidden layer contains LSTM memory blocks. In fact, in the current discussion, the hidden layer is constituted only of these memory blocks. The main element of the LSTM is the memory block: a memory block is a set of memory cells and two gates (see Fig. 6.15). The gates are called input and output gates. Each memory cell (see Fig. 6.16) is composed of a central linear element called the CEC (Constant Error Carousel), two multiplicative units that are controlled by the block's input and output gates, and two non-linear functions $g(\cdot)$ and $h(\cdot)$. The CEC is responsible for keeping the error unchanged for an arbitrarily long time lapse. The multiplicative units controlled by the gates decide when the error should be updated. An LSTM network can have an arbitrary number of memory blocks and each block may have an arbitrary number of cells. The input layer is connected to all the gates and to all the cells. The gates and the cells have input connections from all cells and all gates.

The LSTM topology is represented by $(d : a : b \ (c) : e)$ where d is the number of input features, a the number of neurons in the hidden layer, b the number of memory blocks, c is a comma separated list of the number of cells in each block and e is the number of output layer neurons.

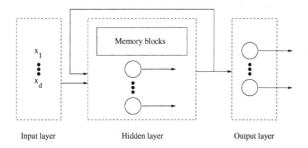

Fig. 6.14 LSTM network overview: the hidden layer can have LSTM memory blocks and simple neurons, although in this section only LSTM memory blocks are used.

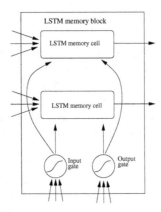

Fig. 6.15 An LSTM memory block with two memory cells and two gates.

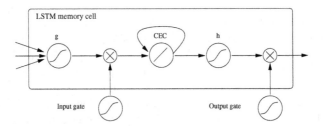

Fig. 6.16 An LSTM memory cell (also visible are the input and output gates).

6.2.2.2 MEE for LSTM

Let the error $E_k = T_k - Y_k$ represent the difference between the target T of the k output neuron and its output Y, at a given time τ (not explicitly given to keep the notation simple). The MMSE of the variable E_k will be replaced by its MEE counterpart. First it is necessary to estimate the PDF of the error. For this purpose, the Parzen window approach is used (see Appendix E) allowing to write the error PDF estimate as:

$$\hat{f}(e_k) = \frac{1}{nh} \sum_{i=1}^{n} K\left(\frac{e_k - e_i}{h}\right) , \qquad (6.23)$$

where, as in expression (3.2), h represents the bandwidth of the kernel K and n is the number of neurons in the output layer.

Using (6.23) one then applies the $\hat{H}_{R_2}$ estimate (F.9) to the MEE approach

$$\hat{H}_{R_2}(e(j)) = -\log\left(\frac{1}{n^2 h} \sum_{i=1}^{n} \sum_{u=1}^{n} K\left(\frac{e(i) - e(u)}{h}\right)\right) . \qquad (6.24)$$

The gradient-based learning algorithm of LSTM presented in [103] can be modified such that the empirical MMSE risk functional is replaced by $\hat{H}_{R_2}$. The change in the derivation presented in [103] occurs in the following expression (the backpropagation error seen at the output neuron k):

$$E_k = f'(net_k)(T_k - Y_k) , \tag{6.25}$$

where $f(\cdot)$ is the sigmoid transfer function, net_k is the activation of the output neuron k at time τ, T_k is the target variable for the output neuron k at time τ and Y_k is the output of neuron k at time τ. The term $(T_k - Y_k)$ in equation (6.25) comes from the derivative of the MSE, $\frac{1}{n}\sum_{i=1}^{n}(t_i - y_i)^2$, w.r.t. the output y_k. This same derivative is computed now, using expression (6.24). Note that, since the logarithm in (6.24) is a monotonically increasing function, to minimize it is the same as to minimize its operand. So, the partial derivative of the operand will be derived, which is

$$\frac{1}{n^2 h \sqrt{2\pi}} \sum_{i=1}^{n} \sum_{j=1}^{n} \frac{\partial}{\partial y_k} \exp\left(-\frac{(e_i - e_j)^2}{2h^2}\right) =$$

$$= \frac{1}{n^2 h \sqrt{2\pi}} \sum_{i=1}^{n} \sum_{j=1}^{n} \frac{\partial}{\partial y_k} \exp\left(-\frac{(t_i - y_i - t_j + y_j)^2}{2h^2}\right) . \tag{6.26}$$

Now, when $i = k$ the derivative of the term inside the summation becomes

$$\exp\left(-\frac{(t_k - y_k - t_j + y_j)^2}{2h^2}\right)\left(-\frac{1}{2h^2}\right) 2(t_k - y_k - t_j + y_j)(-1) . \tag{6.27}$$

Likewise, if $j = k$, the derivative becomes

$$\exp\left(-\frac{(t_i - y_i - t_k + y_k^2}{2h^2}\right)\left(-\frac{1}{2h^2}\right) 2(t_i - y_i - t_k + y_k) . \tag{6.28}$$

Expressions (6.27) and (6.28) yield the same values allowing writing the derivative of the operand of (6.24) as

$$Q \sum_{i=1}^{n} \exp\left(-\frac{(t_i - y_i - t_k + y_k)^2}{2h^2}\right) (t_i - y_i - t_k + y_k) , \tag{6.29}$$

where

$$Q = \frac{2}{n^2 h^3 \sqrt{2\pi}} .$$

So expression (6.25) becomes

$$E_k = Q f'_k(net_k) \sum_{i=1}^{n} \exp\left(-\frac{a_{ik}}{2h^2}\right) a_{ik} , \tag{6.30}$$

where Q stands for a constant term and a_{ik} is

$$a_{ik} = t_i - y_i - t_k + y_k = e_i - e_k \ .$$

6.2.2.3 Experiments

In this section several experiments are presented that compare the performance of LSTM learning with MMSE versus MEE (using R_2EE). Three standard datasets are used for this purpose: the Reber grammar problem, the embedded Reber grammar and the $A^n B^n$ grammar (for an introduction to formal languages see [91]).

The finite-state machine in Fig. 6.17 generates strings from the grammar known as the Reber grammar. The strings are generated by starting at B,

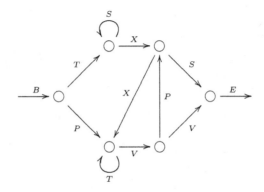

Fig. 6.17 A finite-state machine for the Reber grammar.

and moving from node to node, adding the symbols in the arcs to the string. The symbol E marks the end of a string. When there are two arcs leaving a node, one is chosen randomly with equal probability. This process generates strings of arbitrary length. Several experiments were conducted whose goal was the prediction of the next valid symbol of a string, after the presentation of a given symbol. For instance, if the network receives a starting symbol B it has to predict that the possible next symbols are P or T. If the network is able to correctly predict at any step all possible symbols of all strings generated by the grammar, in the training and test sets, using less than 250 000 sequences for learning, it is considered that the learning has converged. This number is equal to the number of sequences in the dataset times the number of training epochs. So, in our case, since the datasets used have 500 sequences, the number 250 000 is reached with 500 epochs. This number 250 000 was used in [6] to allow a number of sequences large enough for most networks to converge, at the same time avoiding a large computational cost while running the experiments.

Regarding the nature of the problem, although it is a sequence prediction task, we are entitled to regard it too as a classification task at least for strings reaching the terminal node, since a decision can then be made as to the successful parsing of the string.

In the mentioned work [6], a set with 500 strings from the Reber grammar was used for training and a different set with 500 strings for testing. For each LSTM topology and value of the parameter h the training-test process was repeated 100 times, each time with random initialization of the network weights. The strings were coded in a 1-out-of-7 coding, so the number of input features and the number of output layer neurons were both 7. Two topologies were tested: in the first case two memory blocks were used, one with one cell and the other with two cells (Table 6.9); in the second case, both blocks had two cells (Table 6.10). Tables 6.9 and 6.10 show the percentage of the trained networks that were able to perfectly learn both the training and test sets, and the average and standard deviation of the number of sequences that were used for training. Both tables present results for learning rates (η) of 0.1, 0.2 and 0.3. The MMSE lines refers to the use of the original MMSE learning algorithm. The results are discussed bellow.

Table 6.9 Results for the experiments with the Reber grammar using the topology (7:0:2(2,1):7). ANS stands for Average Number of Sequences necessary to converge.

	$\eta = 0.1$		$\eta = 0.2$		$\eta = 0.3$	
	ANS (std) [10^3]	% conv.	ANS (std) [10^3]	% conv.	ANS (std) [10^3]	% conv.
MMSE	15.1 (24.5)	38	74.9 (116.5)	63	61.0 (111.5)	**56**
MEE h=1.3	81.8 (115.4)	36	42.6 (51.5)	11	113.6 (135.6)	7
MEE h=1.4	45.6 (68.2)	45	70.6 (93.8)	11	61.8 (63.6)	10
MEE h=1.5	26.0 (43.9)	54	84.1 (120.2)	29	47.2 (39.1)	13
MEE h=1.6	28.4 (43.1)	**66**	58.0 (84.2)	37	135.1 (160.1)	15
MEE h=1.7	23.0 (25.9)	64	54.9 (87.8)	40	96.9 (135.8)	30
MEE h=1.8	75.8 (51.8)	30	60.1 (96.8)	50	66.0 (111.8)	33
MEE h=1.9	78.0 (110.1)	62	53.6 (94.1)	61	48.7 (66.2)	33
MEE h=2.0	49.3 (77.6)	58	57.6 (109.0)	**67**	57.4 (83.7)	51

The second set of experiments used strings from the embedded Reber Grammar (ERG) generated as shown in Fig. 6.18. This grammar produces two types of strings: BT <Reber string> TE and BP <Reber string> PE. In order to recognize these strings, the learning machine has to be able to distinguish them from strings such as BP <Reber string> TE and BT<Reber string>PE. To do this it is essential to remember the second symbol in the sequence such that it can be compared with the second last symbol. Notice that the length of the sequence can be arbitrarily large. This problem is no longer learnable by an Elman net and a RTRL net only learns

Table 6.10 Results for the experiments with the Reber grammar using the topology (7:0:2(2,2):7). ANS stands for Average Number of Sequences necessary to converge.

	$\eta = 0.1$		$\eta = 0.2$		$\eta = 0.3$	
	ANS (std) [10^3]	% conv.	ANS (std) [10^3]	% conv.	ANS (std) [10^3]	% conv.
MMSE	19.2 (26.1)	41	46.0 (77.2)	53	30.1 (50.7)	61
MEE h=1.3	57.2 (67.5)	43	44.3 (58.8)	18	28.4 (12.9)	5
MEE h=1.4	20.2 (27.8)	55	70.5 (112.4)	22	108.9 (130.5)	10
MEE h=1.5	33.6 (55.3)	59	68.1 (109.8)	36	31.7 (32.7)	19
MEE h=1.6	26.5 (41.9)	**65**	38.3 (47.2)	42	58.4 (82.7)	25
MEE h=1.7	32.3 (57.2)	53	48.4 (83.8)	48	55.6 (82.9)	26
MEE h=1.8	24.5 (48.8)	60	54.4 (88.3)	61	43.4 (82.0)	40
MEE h=1.9	51.8 (76.7)	49	70.6 (134.5)	66	62.1 (108.1)	**66**
MEE h=2.0	44.6 (79.3)	48	48.3 (85.8)	**68**	44.3 (70.2)	41

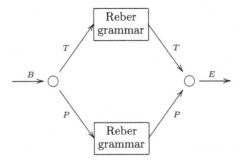

Fig. 6.18 A finite-state machine for the embedded Reber grammar.

it with some difficulty, since, as opposed to the Reber grammar problem, there is the need to retain information for long time lapses.

In this case the experiments reported in [6] were similar to the ones described for the Reber grammar but the learning rates used were 0.1 and 0.3. The train and test sets had also 500 strings each. The topology used was 7:0:4(3,3,3,3):7 (the coding is the same as in the Reber grammar example). The experiments were also repeated 100 times. The results are in Table 6.11.

The last dataset consists of a series of strings from the grammar $A^n B^n$. Valid strings consist of n A symbols followed by exactly n B symbols. The network is trained with only correct strings, and n from 1 up to 10. It is considered that the network converged if it is able to correctly classify all the strings in both training and test sets, using less than 50 000 sequences for learning. In the first experiment reported in [6] the correct strings were used for the test set for $n = 1 \ldots 50$. In the second experiment the correct strings were used for $n = 1 \ldots 100$. In both experiments the network topology was 3:0:2(1,1):3. Both experiments were repeated 100 times for $\eta = 1$. The results obtained are in Table 6.12.

Table 6.11 Results for the experiments with the embedded Reber grammar. ANS stands for Average Number of Sequences necessary to converge.

	$\eta = 0.1$		$\eta = 0.3$	
	ANS (std) $[10^3]$	% conv.	ANS (std) $[10^3]$	% conv.
MMSE	44.4(48.4)	8	66.3 (40.1)	6
MEE h=1.3	79.6 (94.5)	7	-	0
MEE h=1.4	13.1 (6.5)	8	-	0
MEE h=1.5	49.9 (53.8)	12	327 (-)	1
MEE h=1.6	65.7 (46.9)	8	62.0 (74.5)	3
MEE h=1.7	41.2 (26.9)	8	183.7 (137.4)	3
MEE h=1.8	60.6 (35.8)	12	102.4 (101.4)	5
MEE h=1.9	124.3 62.4	13	76.9 (125.0)	**7**
MEE h=2.0	143.2 (111.9)	**14**	84.6 (87.6)	**7**

Table 6.12 Results for the experiments with the grammar $A^n B^n$. ANS stands for Average Number of Sequences necessary to converge.

	Test $n = 1, \dots, 50$		Test $n = 1, \dots, 100$	
	Average n. seq. (std) $[10^3]$	% conv.	Average n. seq. (std) $[10^3]$	% conv.
MMSE	4.93 (2.80)	17	4.90 (2.41)	12
MEE h=1.3	10.31 (7.40)	18	8.75 (6.48)	13
MEE h=1.4	11.67 (8.11)	19	11.90 (7.94)	12
MEE h=1.5	15.31 (8.25)	28	16.08 (7.76)	18
MEE h=1.6	17.06 (8.62)	36	17.51 (8.46)	25
MEE h=1.7	18.77 (8.78)	45	18.33 (8.33)	29
MEE h=1.8	20.74 (8.74)	**50**	19.69 (7.83)	**35**
MEE h=1.9	20.80 (8.38)	48	21.09 (7.39)	**35**
MEE h=2.0	22.19 (8.20)	49	22.50 (7.52)	33

6.2.2.4 Discussion

In the Reber grammar experiments, the MEE approach improved the percentage of convergence in all six sets of experiments except for the first topology with $\eta = 0.3$. In one case (first topology and $\eta = 0.2$) it not only outperformed MMSE but there was also a statistical significant average reduction in the number of sequences necessary for the network to converge from 74.9×10^3 to 57.6×10^3 (t-test $p < 0.01$).

In the case of the experiments with the ERG, and for $\eta = 0.1$, two benefits were found from the application of MEE: in two cases ($h = 1.4$ and 1.7) the same percentage of convergence as with MMSE was obtained but with a smaller number of required training sequences on average, highly significant for $h = 1.4$ (t-test $p \approx 0$). In four other cases, the percentage of convergence increased from 8 to 12, 13 and 14%. In these cases the number of necessary sequences also increased when compared to the use of MMSE. When η was set at 0.3, there were two cases ($h = 1.3$ and 1.4) where MEE was not able to converge. Nevertheless the best result was still obtained with MEE for $h = 1.9$ and 2.0. In these experiments it is apparent that the increase of η was neither beneficial for MMSE nor for MEE.

Finally, the experiments with the $A^n B^n$ grammar confirmed that the use of MEE is beneficial in terms of increasing the convergence percentage: from the 16 sets of 100 repetitions only one had the same performance of MMSE; MEE performance improved in all other sets, although at the cost of an increased number of training sequences.

6.3 Complex-Valued Neural Network

6.3.1 Introduction

Complex-valued neural networks (CVNNs) have been gaining considerable attention in recent years [101, 148, 9, 206]. The benefits of using a complex-valued NN are manifest when dealing with specific types of data, such as wave phenomena [100, 101] where there is a need or convenience to process phase and amplitude information as a single object.

The key feature of these networks is related to how the product of complex numbers works. Let's compare what happens when a two-dimensional input to a neuron is considered in the following two cases: first, the traditional real-valued case where the neuron has a weight associated with each input; second the complex value case where a single complex weight is used for a single complex-valued input. The two cases are represented in Fig. 6.19. Consider the real numbers a, b, w_1 and w_2. In a real-value neuron, the input values a and b are multiplied by the respective weights w_1 and w_2 producing an input of $aw_1 + bw_2$ to the neuron. In the neuron represented in the right part of the figure, the same input values a and b are there but now as real and imaginary parts of a complex input $z = a + ib$, with $i = \sqrt{-1}$ representing the imaginary unit. The weights are also part of a single complex weight $w = w_1 + iw_2$. The neuron now sees this input as the product $zw = aw_1 - bw_2 + i(aw_2 + bw_1)$. If this is written out using amplitude and phase representation, $z = \sqrt{a^2 + b^2}e^{i\mathrm{atan}(b/a)}$ and $w = \sqrt{w_1^2 + w_2^2}e^{i\mathrm{atan}(w_2/w_1)}$, it leads to $zw = \sqrt{a^2 + b^2}\sqrt{w_1^2 + w_2^2}e^{i(\mathrm{atan}(b/a)+\mathrm{atan}(w_2/w_1))}$. This means that the product of complex numbers amounts to simply multiplying the amplitudes and adding the phases.

Fig. 6.19 Example of how a real-valued (left) and a complex-valued (right) neuron deal with a two-dimensional input. $z = a + ib$ and $w = w_1 + iw_2$. See the text for details.

In this section, MMSE-batch and MEE-based learning algorithms for the single layer CVNN, with the architecture proposed in [9], are presented.

6.3.2 Single Layer Complex-Valued NN

This subsection follows closely [9]. Consider an input space with d features. The neuron input is the complex vector $x = x^R + ix^I$ where x^R is the real and x^I is the imaginary part, such that $x \in \mathbb{C}^d$. The weight matrix $w \in \mathbb{C}^d$ can also be written as $w = w^R + iw^I$. The net input of neuron k is given by

$$z_k = \theta_k + \sum_{j=1}^{d} w_{kj} x_j \ , \tag{6.31}$$

where $\theta_k \in \mathbb{C}$ is the bias for neuron k and can be written as $\theta_k = \theta_k^R + i\theta_k^I$. Given the complex multiplication of $w_k x$, this can further be written as

$$z_k = z_k^R + iz_k^I = \left(\theta_k^R + x^R w_k^R - x^I w_k^I\right) + i\left(\theta_k^I + x^R w_k^I + x^I w_k^R\right) \ . \tag{6.32}$$

Note that,

$$x^R w_k^R = \sum_{j=1}^{m} x_j^R w_{kj}^R \ . \tag{6.33}$$

The k neuron output is given by $y_k = f(z_k)$ where $f : \mathbb{C} \to \mathbb{R}$ is the activation function and $y_k \in \mathbb{R}$. The activation function used is $f(z_k) = (s(z_k^R) - s(z_k^I))^2$ where $s(\cdot)$ is the sigmoid function $s(x) = \frac{1}{1+\exp(-x)}$. Given the form of this activation function, it is possible to solve non-linear classification problems, whereas in the case of a real-valued neural network with only one layer (such as a simple perceptron) this would not be possible. Now that the output of the single complex-valued neuron is defined, the way to train a network composed of a single layer with N of these complex-valued neurons can be presented. The network is trained in [9] by applying gradient descent to the MMSE functional

$$E(w) = \frac{1}{2} \sum_{k=1}^{N} (t_k - y_k)^2 \ , \tag{6.34}$$

where $t_k \in \mathbb{R}$ represents the target output for neuron k. To minimize (6.34) the derivative w.r.t. the weights is used:

$$\frac{\partial E}{\partial w_{kj}^R} = -2e_k(s(z_k^R) - s(z_k^I))\left(s'(z_k^R)x_j^R - s'(z_k^I)x_j^I\right) \ . \tag{6.35}$$

The previous expression is the derivative w.r.t. the real weights but a similar one should be made w.r.t. the imaginary weights. To obtain the weights, the gradient descent rule is used, with the update of the weights at each iteration (m), that is, after the presentation of each training pattern to the network, written as

$$w_{kj}^{R,(m)} = w_{kj}^{R,(m-1)} - \Delta w_{kj}^{R,(m)} \; , \qquad (6.36)$$

with

$$\Delta w_{kj}^{R,(m)} = \eta \frac{\partial E}{\partial w_{kj}^R}\bigg|_{w_{kj}^{R,(m)}} = 2\eta e_k (s(z_k^R) - s(z_k^I)) \left(s'(z_k^R)x_j^R - s'(z_k^I)x_j^I \right) \; . \tag{6.37}$$

A similar derivation can be made for the case of the imaginary part of the weights, yielding

$$\Delta w_{kj}^{I,(m)} = \eta \frac{\partial E}{\partial w_{kj}^I}\bigg|_{w_{kj}^{I,(m)}} = 2\eta e_k (s(z_k^I) - s(z_k^R)) \left(s'(z_k^R)x_j^I + s'(z_k^I)x_j^R \right) \; . \tag{6.38}$$

It is possible to show that the final expressions for the adjustment of the real and imaginary parts of the bias are

$$\Delta\theta_k^{R,(m)} = 2\eta e_k (s(z_k^R) - s(z_k^I))s'(z_k^R) \tag{6.39}$$

and

$$\Delta\theta_k^{I,(m)} = 2\eta e_k (s(z_k^I) - s(z_k^R))s'(z_k^I) \; . \tag{6.40}$$

6.3.3 MMSE Batch Algorithm

This section presents the batch version of the algorithm studied in the preceding section, as proposed in [3]. The change w.r.t. the original algorithm is on the empirical risk functional to be minimized: instead of (6.34) as in the original algorithm [9] it now contains the error contributions from all n patterns in the training set,

$$E(w) = \frac{1}{2L} \sum_{l=1}^{n} \sum_{k=1}^{N} (t_k - y_k)^2 \; . \tag{6.41}$$

The difference between this batch approach and the stochastic one presented earlier is, as usual, that the values of Δw_{kj} and $\Delta\theta_k$ obtained after each pattern is presented to the network are summed and the weights are only updated at the end of each epoch.

6.3.4   MEE Algorithm

The minimum error entropy (MEE) will now be used instead of the minimum mean square error (MMSE) as the optimization principle behind the learning algorithm for this network. MEE training needs a batch mode algorithm because the distribution of the errors for updating the weights has to be estimated, so several of these errors are needed to obtain a good PDF estimate.

As seen above, the error $e_j = t_j - y_j$ represents the difference between the target t_j of the j neuron and its output y_j. The MSE of the variable e_j will be replaced with its EE counterpart. First it is necessary to estimate the PDF of the error. For this, the Parzen window estimator is used.

Using the empirical estimate of H_2EE, as in expression (3.5), we minimize the corresponding information potential, which (ignoring constant factors) is written as

$$V_{R_2} = \sum_{i=1}^{n} \sum_{u=1}^{n} K\left(\frac{e_i - e_u}{h}\right) . \tag{6.42}$$

The derivative of V_{R_2} w.r.t. the real weights is:

$$\frac{\partial V_{R_2}}{\partial w_{kj}^R} = \frac{1}{h} \sum_{i=1}^{n} \sum_{u=1}^{n} K'\left(\frac{e_i - e_u}{h}\right)\left(\frac{\partial e_i}{\partial w_{kj}^R} - \frac{\partial e_u}{\partial w_{kj}^R}\right) . \tag{6.43}$$

Since $\frac{\partial e_i}{\partial w_{kj}^R}$ is given by $-\frac{\partial y_i}{\partial w_{kj}^R}$, expression (6.43) can be written as

$$\frac{\partial V_{R_2}}{\partial w_{kj}^R} = \frac{2}{h} \sum_{i=1}^{n} \sum_{u=1}^{n} K'\left(\frac{e_i - e_u}{h}\right)((s(z_i^R) - s(z_i^I))(s'(z_i^R)x_j^R - s'(z_i^I)x_j^I)-$$
$$-(s(z_u^R) - s(z_u^I))(s'(z_u^R)x_j^R - s'(z_u^I)x_j^I)) . \tag{6.44}$$

A gradient ascent procedure is used instead of gradient descent since the goal is to maximize V_{R_2}. So, the weight update at each iteration (m) is guided by

$$\Delta w_{kj}^R(m) = \eta \frac{\partial V_{R_2}}{\partial w_{kj}^R} . \tag{6.45}$$

A similar derivation can be done for the the imaginary weights. The expression equivalent to (6.44) is

$$\frac{\partial V_{R_2}}{\partial w_{kj}^I} = \frac{2}{h} \sum_{i=1}^{n} \sum_{u=1}^{n} K'\left(\frac{e_i - e_u}{h}\right)((s(z_i^I) - s(z_i^R))(s'(z_i^R)x_j^I + s'(z_i^I)x_j^R)-$$
$$-(s(z_u^I) - s(z_u^R))(s'(z_u^R)x_j^I + s'(z_u^I)x_j^R)) . \tag{6.46}$$

The update equations for the thresholds can be obtained by finding $\frac{\partial V_{R_2}}{\partial \theta_k^R}$ and $\frac{\partial V_{R_2}}{\partial \theta_k^I}$. These equations are

$$\frac{\partial V_{R_2}}{\partial \theta_k^R} = \frac{2}{h} \sum_{i=1}^{n} \sum_{u=1}^{n} K' \left(\frac{e_i - e_u}{h} \right) ((s(z_i^R) - s(z_i^I))s'(z_i^R) -$$
$$- (s(z_u^R) - s(z_u^I))s'(z_u^R)) \tag{6.47}$$

and

$$\frac{\partial V_{R_2}}{\partial \theta_k^I} = \frac{2}{h} \sum_{i=1}^{n} \sum_{u=1}^{n} K' \left(\frac{e_i - e_u}{h} \right) ((s(z_i^I) - s(z_i^R))s'(z_i^I) -$$
$$- (s(z_u^I) - s(z_u^R))s'(z_u^I)) \, . \tag{6.48}$$

6.3.5 Experiments

In the following we describe experiments presented in [3] where the original single layer CVNN is compared against other classifiers, including the MEE-based (R_2EE) version of the single layer CVNN.

6.3.5.1 Datasets

An artificially generated dataset (Checkerboard) as in Fig. 6.10, was used in [3] which consisted of a 2 by 2 grid of domains with alternate classes in the domains. The dataset had 400 instances (100 per grid position and 200 per class). In this case it is considered that the value of the X coordinate of a point is the real part of a complex measurement and the Y coordinate is the imaginary part. The second dataset is the breast cancer dataset studied in [211]. It consists of electrical impedance measurements that were performed on 120 samples of freshly excised breast tissue. The dataset has 6 classes, 120 instances and 24 features (real and imaginary parts of 12 measurements of impedance at different frequencies). The data was normalized with zero mean and unit standard deviation for all algorithms with the exception of the SVM where a normalization in the interval $[-1, 1]$ was done for each feature.

6.3.5.2 Results

The results are in Table 6.13 showing average errors and standard deviations of 30 repetitions of two-fold cross-validation. Besides the results obtained with MMSE and MEE, Table 6.13 also shows the results obtained using SVM with

RBF kernel [227] (best value for parameter g varying from 2.2 to 0.8 in steps
of 0.2, for $C = 10$ and $C = 100$), k-NN [59] (best value for k=1, 3, 5 and 7)
and the C4.5 decision tree [177]. For the MEE version three results for each
value of the learning rate were obtained for kernel bandwidths of 1.0, 1.2 and
1.4. Only the best results are reported in Table 6.13. Also the results shown
for the CVNNs were the best obtained when training ran for 4 000 epochs,
evaluated at 20 epochs intervals on the test set.

The results for the Checkerboard problem are impressive: the CVNN is
able to attain almost perfect classification with the second best method, SVM
with RBF, lagging behind. The different performance is statistically highly
significant: e.g., t-test $p \approx 0$ for the last row MEE average error rate of 0.23.
For this dataset, the MEE algorithm is also the best one for the tested values
of the parameters, when compared with the other two CVNN versions; even
when one takes the $\eta = 0.07$ case, where the difference between the MEE
and the MMSE algorithms is the smallest, the superiorness of the MEE is
statistically significant (t-test $p = 0.006$).

For the Brest Cancer problem, the SVM with RBF was the best classifier.
The CVNN algorithms came in second place. However the difference between
the SVM performance and the best MEE performance (33.00) has no statistical significance (t-test $p = 0.148$). The same can be said of the difference
between the MEE CVNN and the other CVNNs.

Table 6.13 Average error in percentage, with standard deviation for 30 repetitions
of a two fold-cross validation for both datasets. CVNN B-MMSE is the batch MMSE
algorithm.

| Classifier | Checkerboard | | Breast cancer | |
	Parameters	Error (std)	Parameters	Error (std)
SVM RBF	g=1.8, C=10	2.92 (0.60)	g=1.0, C=10	31.83 (3.23)
k-NN	$k = 1$	4.48 (0.98)	$k = 5$	34.42 (2.99)
C4.5	None	25.22 (0.29)	None	35.28 (5.28)
CVNN MMSE	$\eta = 0.09$	0.51 (0.38)	$\eta = 0.09$	32.11 (5.68)
CVNN B-MMSE	$\eta = 0.09$	0.60 (0.43)	$\eta = 0.09$	33.25 (6.05)
CVNN MEE	$\eta = 0.09, h = 1.0$	0.30 (0.22)	$\eta = 0.09, h = 1.0$	33.14 (5.55)
CVNN MMSE	$\eta = 0.07$	0.43 (0.26)	$\eta = 0.07$	32.69 (5.30)
CVNN B-MMSE	$\eta = 0.07$	0.48 (0.39)	$\eta = 0.07$	33.25 (6.05)
CVNN MEE	$\eta = 0.07, h = 1.4$	0.32 (0.31)	$\eta = 0.07, h = 1.4$	33.47 (6.19)
CVNN MMSE	$\eta = 0.05$	0.57 (0.50)	$\eta = 0.05$	33.64 (5.07)
CVNN B-MMSE	$\eta = 0.05$	0.45 (0.30)	$\eta = 0.05$	33.03 (5.26)
CVNN MEE	$\eta = 0.05, h = 1.4$	0.33 (0.24)	$\eta = 0.05, h = 1.0$	33.00 (4.94)
CVNN MMSE	$\eta = 0.03$	0.72 (0.55)	$\eta = 0.03$	33.17 (6.18)
CVNN B-MMSE	$\eta = 0.03$	0.58 (0.36)	$\eta = 0.03$	32.94 (5.74)
CVNN MEE	$\eta = 0.03, h = 1.4$	0.37 (0.22)	$\eta = 0.03, h = 1.0$	33.50 (4.43)
CVNN MMSE	$\eta = 0.01$	0.68 (0.32)	$\eta = 0.01$	33.28 (5.66)
CVNN B-MMSE	$\eta = 0.01$	0.68 (0.54)	$\eta = 0.01$	33.61 (5.29)
CVNN MEE	$\eta = 0.01, h = 1.0$	0.23 (0.31)	$\eta = 0.01, h = 1.0$	34.58 (3.84)

6.4 An Entropic Clustering Algorithm

6.4.1 Introduction

Clustering is an everyday activity. People group objects into categories, classes, types or other structures based on their similitude or differences. Clustering deals with the process of forming or finding different groups in a given set, based on similarities or dissimilarities among their objects. How can one define the similarities or the dissimilarities?; how many groups are there in a given set?; how do groups differ from each other or how can we find them? These are examples of some basic questions, none with a unique answer.

Let X_{ds} be a dataset, $X_{ds} = \{\mathbf{x}_i\}$, $i = 1, 2, ..., n$, where n is the number of instances (objects, points) and $\mathbf{x}_i$ a d-dimensional vector representing each instance. We define a c-clustering of X_{ds} as a partition of X_{ds} into c clusters (sets) $C_1, C_2, ..., C_c$, obeying the conditions: $C_i \neq \oslash$, $i = 1, ..., c$; $\cup_{i=1}^{c} C_i = X_{ds}$; $C_i \cap C_j = \oslash$, $i \neq j$, $i, j = 1, ..., c$; $x \in C_i$ iif $d(\{\mathbf{x}_i\}, C_i) \leq d(\{\mathbf{x}_i\}, C_j), \forall i \neq j = 1, ..., c$, where $d(\{\mathbf{x}_i\}, S)$ denotes the dissimilarity value of the singleton set $\{\mathbf{x}\}$ w.r.t. any set $S \subset X_{ds}$. Each instance given these conditions, belongs to a single cluster. This kind of clustering is known as *hard clustering* (there are algorithms, like those based on fuzzy theory, in which an element has degrees of membership for each cluster: soft clustering.)

There is a wide variety of techniques to perform clustering. Results are not unique; they depend on the evaluation of dissimilarity and on the algorithm used to perform it. The same data can be clustered with different acceptable solutions depending on the clustering algorithm. Hierarchical (tree) clustering, for example, gives several solutions according to the tree level chosen for the final solution.

There are algorithms based on stepwise merging or division of subsets, known as sequential or hierarchical algorithms; others, are based on the principle of function approximation, like fuzzy clustering or density based algorithms; yet others, are based on graph theory or competitive learning. We present here an entropic dissimilarity measure and a combination of hierarchical and graph approaches to perform clustering. The respective algorithm, LEGClust (Layered Entropic subGraph Clustering), is sensitive to the local structure of the data, achieving clusters that reflect the local inner structure. Moreover, we shall see that LEGClust reflects the MEE concept in some adequate way.

6.4.2   Overview of Clustering Algorithms

We present a brief overview of some clustering algorithms, specially those used in the comparative experiments presented later. Further details on algorithms, techniques and applications, can be found in [247, 112, 113, 114, 25, 179, 225]).

From the huge variety of clustering algorithms, the hierarchical agglomerative ones are probably the most used even today. Hierarchical agglomerative algorithms create, by definition, a hierarchy of clusters from the dataset. They start by assigning each instance to a single cluster and then, based on dissimilarity measures, proceed to merge small clusters into larger ones in a stepwise manner. The process ends when all instances in the dataset are members of a single cluster. The resulting hierarchical tree defines the clustering levels. Examples of hierarchical clustering algorithms are CURE [85], ROCK [86], AGNES [128], BIRCH [253, 254] and Chameleon [127].

Other clustering algorithms like those based on graphs and graph theory were proposed in the early 1970's. They use the high connectivity in similarity graphs to perform clustering [154, 155]. These algorithms are usually divisive algorithms, meaning that they first consider a highly connected graph (that corresponds to a single cluster) containing all elements of the dataset and then start doing consecutive cuts in the graph to obtain smaller clusters. A cut in a graph corresponds to the removal of a set of edges that disconnects the graph. A minimum cut (min-cut) is the removal of the smallest number of edges that produces a cut. The result of a cut in the graph causes the splitting of one cluster into, at least, two clusters. Examples of graph-based clustering algorithms can be found in [120, 92, 93, 244]. Chameleon [127], mentioned earlier, also uses a graph-theoretic approach.

Graph cutting is also used in spectral clustering, commonly applied in image segmentation and, more recently, in web page and document clustering and also in bioinformatics. The rationale of spectral clustering is to use the special properties of the eigenvectors of a Laplacian matrix as the basis to perform clustering. Fidler [73] was one of the first to show the application of eigenvectors to graph partitioning. The Laplacian matrix is based on an affinity matrix built with a similarity measure. The most common similarity measure used in spectral clustering is $\exp\left(-d_{ij}^2/2\sigma^2\right)$, where d_{ij} is the Euclidian distance between data vectors $\mathbf{x}_i$ and $\mathbf{x}_j$ and σ is a scaling parameter.

Several spectral clustering algorithms differ in the way they use the eigenvectors in order to perform clustering. Some researchers use the eigenvectors of the "normalized" Laplacian matrix [42] (or a similar one) in order to perform the cutting, usually using the second smallest eigenvector [209, 124, 55]. Others, use the highest eigenvectors as input to other clustering algorithm [166, 157]. A comparison of several spectral clustering algorithms can be found in [229].

Other clustering algorithms use the different density regions of the data to perform clustering. Examples of these density-based clustering algorithms are

DBScan [69] and Mean Shift [77, 40, 44, 45] algorithms. Mean Shift has some very good results in image segmentation and computer vision applications.

In a later section we will present experimental results comparing clustering algorithms of the above types with LEGClust.

6.4.2.1 Clustering with Entropy

Clustering algorithms applying concepts of entropy, mutual information and the Kullback-Leibler divergence, have been proposed by several authors. Examples of such algorithms are the minimum entropic clustering [137], entropic spanning graphs clustering [98] and entropic subspace clustering [39]. In other works entropy is used as a measure of proximity or interrelation between clusters. Examples of these algorithms are those proposed by Jenssen [117] and Gokcay [84], that use a so-called Between-Cluster Entropy, and the one proposed by Lee [134, 135] that uses the Within-Cluster Association. A common characteristic of these algorithms is that they start by selecting random seeds for the first clusters which may produce very different results in the final clustering solution. Recent works by He [97], Vretos [234] and Faivishevsky [71] are examples of clustering algorithms using mutual information.

The main characteristics of these entropy-based algorithms are their high complexity and that they are heavily time consuming. They are, nonetheless, attractive because they present some good results in several specific applications.

The LEGClust algorithm described in the following sections uses Renyi's quadratic entropy, H_{R_2}, which was seen in the previous chapters to be computationally simpler to use than H_S or H_{R_α}, $\alpha \neq 2$. One could however use other entropic measures as well.

Renyi's quadratic entropy was already discussed earlier in this book. LEGClust uses the multivariate version of formula (F.10)

$$\hat{H}_{R_2} = -\ln\left(\frac{1}{N^2}\sum_{i=1}^{N}\sum_{j=1}^{N} g(\mathbf{x}_i - \mathbf{x}_j; \mathbf{0}, 2h^2\mathbf{I})\right), \qquad (6.49)$$

where $\mathbf{x}_i$ and $\mathbf{x}_j$ are the data vectors.

6.4.3 Dissimilarity Matrix

As stated earlier, clustering algorithms are based on similarity or dissimilarity measures between the objects (data instances) of a set.

Objects belonging to the same partition or cluster possess a higher degree of similarity with each other than with any other objects from other clusters.

This degree of similarity is defined by a similarity or dissimilarity measure, where measure is allowed to be interpreted here in a lax sense, as some monotonic set function $d(C_i, C_j)$ whose minimum value is $d(C_i, C_i)$ for any C_i.

The most common dissimilarity measure between two real-valued vectors $\mathbf{x}$ and $\mathbf{y}$, is the weighted L_p metric,

$$d_p(\mathbf{x}, \mathbf{y}) = \left(\sum_{i=1}^{d} w_i |x_i - y_i|^p \right)^{\frac{1}{p}}, \qquad (6.50)$$

where x_i and y_i are the ith coordinates of $\mathbf{x}$ and $\mathbf{y}$, $i = 1, ..., d$, and $w_i \geq 0$ is the ith weight coefficient. The unweighted ($\mathbf{w} = 1$) L_p metric is also known as Minkowski distance of order p ($p \geq 1$). Examples of this distance are the well-known Euclidian distance — the most common dissimilarity measure used by clustering algorithms —, obtained by setting $p = 2$, the Manhattan distance, for $p = 1$, and the L_∞ or Chebyshev distance. The LEGClust algorithm also uses a dissimilarity measure, although defined in an unconventional way. Dissimilarities between objects $\mathbf{x}_i$ and $\mathbf{x}_j$, for all objects represented by a set of vectors $\{\mathbf{x}_1, \mathbf{x}_2, \ldots, \mathbf{x}_n\}, \mathbf{x}_i \in \mathbb{R}^d$, are conveniently arranged in a *dissimilarity matrix* $\mathbf{A} \in \mathbb{R}^{n \times n}$, where each element of $\mathbf{A}$ is $a_{i,j} = d(\mathbf{x}_i, \mathbf{x}_j)$ with $d(\mathbf{x}_i, \mathbf{x}_j)$ the dissimilarity between $\mathbf{x}_i$ and $\mathbf{x}_j$ (in rigor, $d(\{\mathbf{x}_i\}, \{\mathbf{x}_j\})$) for the singleton sets $\{\mathbf{x}_i\}$ and $\{\mathbf{x}_j\}$).

6.4.4 The LEGClust Algorithm

As mentioned earlier, clustering solutions may vary widely with the algorithm being used and, for the same algorithm, with its specific settings. People also cluster data differently according to their knowledge, perspective or experience. In [201] some clustering tests were performed involving several types of individuals in order to try to understand the mental process of data clustering. The tests used two-dimensional datasets similar to those to be presented in Sect. 6.4.5. Figure 6.20 shows one such dataset with different clustering solutions suggested by different individuals.

The most important conclusion presented in [201] was that human clustering exhibits some balance between the importance given to local (e.g., connectedness) and global (e.g., structuring direction) features of the data. The tests also provided majority choices of clustering solutions that one can use to compare the results of different clustering algorithms.

The following sections describe the LEGClust algorithm, first presented in [204]. We first introduce the entropic dissimilarity matrix and, based on that, the computation of the so-called layered entropic proximity matrix.

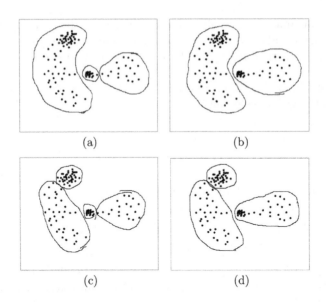

Fig. 6.20 An example of clustering solutions for a particular dataset. Children usually propose solution b) and adults solutions b) and c). Solution d) was never proposed in the study reported in [201].

6.4.4.1 The LEGClust Dissimilarity Matrix

Let us consider the set of objects (points) depicted in Fig. 6.21. These points are in a square grid in two-dimensional[3] space x_1-x_2, except for point Q. Let us denote:

- $K = \{k_i\}$, $i = 1, 2, .., M$, the set of the M nearest neighbors of Q;
- $\mathbf{d}_{ij}$, the difference vector between points k_i and k_j, $\mathbf{d}_{ij} = \mathbf{k}_j - \mathbf{k}_i$ for all $i, j = 1, 2, .., M$, $i \neq j$. These are the *connecting vectors* between those points and there are $M(M-1)$ such vectors;
- $\mathbf{q}_i$, the difference vector between point Q and each of the M-nearest neighbors k_i.

Despite the fact that the shortest connection between Q and one of its neighbors is q_1 we clearly see that candidates for "ideal connection" are those connecting Q with P or with R because they reflect the local structure of the data.

Let us represent all $\mathbf{d}_{ij}$ connecting vectors translated to a common origin as shown in Fig. 6.22. We call this an *M-neighborhood vector field*. Since we have a square grid, there are a lot of equal overlapped vectors.

---

[3] For simplicity we use a two-dimensional dataset, but the analysis is valid for higher dimensions.

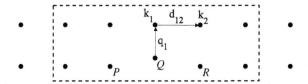

Fig. 6.21 A simple example with M-nearest neighbors of point Q, $M = 9$; the 9-neighborhood of Q corresponds to points inside the dotted region.

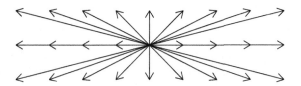

Fig. 6.22 The 9-neighborhood vector field of point Q.

An M-neighborhood vector field can be interpreted as a probability density function and represented, in this example, by a two-dimensional empirical PDF $\mu_M(\mathbf{x}_1, \mathbf{x}_2)$ shown in Fig. 6.23, where each $(\mathbf{x}_1, \mathbf{x}_2)$ pair corresponds to $\mathbf{d}_{ij}$ vector ends. The vertical bars represent $\mu_M(\mathbf{x}_1, \mathbf{x}_2)$, i.e., ocorrence rates of $\mathbf{d}_{ij}$ vector ends. Note that for any selected data point (Q on the present example) every point belonging to some $\mathbb{R}^d$ subset is a candidate vector-end of the M-neighborhood vector field.

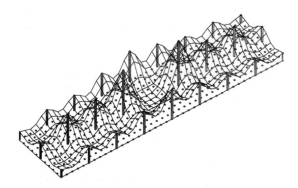

Fig. 6.23 The histogram of the 9-neighborhood vector field of point Q with a mesh showing the respective PDF estimate.

As shown in Fig. 6.23, the probability density function associated with point Q, reflects an horizontal M-neighborhood structure and, therefore, the "ideal connection" for Q should follow this horizontal direction. Besides choosing the ideal connection LEGClust also ranks all possible connections for point Q. It achieves this by comparing, from an entropic point of view, all PDFs resulting from adding each connection vector $\mathbf{q}_i$ to the M-neighborhood vector field. LEGCLust ranks each $\mathbf{q}_i$ according to the variation it introduces into the respective (PDF) entropy. The connection that introduces less disorder into the system, that least increases the entropy of the system, will be top ranked as the *ideal* or *stronger* connection, followed by the other $M - 1$ connections in decreasing order.

Let $D = \{\mathbf{d}_{i,j}\}$, $i, j = 1, 2, .., M$, $i \neq j$. Let $H_{R_2}(D, \mathbf{q}_i)$ be Rényi's quadratic entropy associated with connection $\mathbf{q}_i$, the entropy of the set of all $\mathbf{d}_{ij}$ connections plus connection $\mathbf{q}_i$:

$$H_{R_2}(D, \mathbf{q}_i) = H(D \cup \{\mathbf{q}_i\}), \; i = 1, 2, .., M. \tag{6.51}$$

This entropy is the dissimilarity measure for the LEGClust algorithm.

Let us apply the estimate expressed by formula (6.49) to (6.51). For that purpose we first need a simple and uniform notation for the connection vectors. We do this as follows: $\mathbf{d}_p$, $p = 1, \ldots, M(M - 1)$ will denote any of the $\mathbf{d}_{ij}$ vectors of the M-neighborhood vector field, by setting $p = (i - 1)(M - 1) + j - (i < j)$ for all $i \neq j$; $\mathbf{d}_0$ denotes any particular $\mathbf{q}_i$ connection. Taking the information potential $\hat{V}_{R_2} \equiv \hat{V}_{R_2}(D \cup \{\mathbf{q}_i\})$ (see Appendix F), we have with $n = [M(M - 1)]^2 + M(M - 1)$,

$$\hat{V}_{R_2} = \frac{1}{(2\pi)^{d/2}n^2} \sum_{p=0}^{M(M-1)} \sum_{q=0}^{M(M-1)} e^{-\frac{1}{2}\|\mathbf{d}_p - \mathbf{d}_q\|^2/(2h^2)^d} . \tag{6.52}$$

Interpreting the $\mathbf{d}_p$ (or $\mathbf{d}_q$) as errors, since they represent difference vectors (deviations) between points, one is then searching the $\mathbf{d}_0$ that minimizes Rényi's quadratic entropy of the errors. This bears some resemblance with the MEE approach for supervised classification. Note that any of the previously mentioned entropic clustering algorithms (Sect. 6.4.2.1) simply apply entropy to the data itself. LEGClust is so far the only algorithm that applies a MEE-like concept. The $\hat{V}_{R_2}$ expression can be further processed as

$$\hat{V}_{R_2} = \hat{V}_M + \frac{1}{(2\pi)^{d/2}n^2} \sum_{q=1}^{M(M-1)} e^{-\frac{1}{2}\|\mathbf{d}_0 - \mathbf{d}_q\|^2/(2h^2)^d} , \tag{6.53}$$

where $\hat{V}_M$ is the information potential relative to the $M(M - 1)$ radially symmetric distribution of the $\mathbf{d}_{ij}$. $\hat{V}_M$ is constant for a given M-neighborhood vector field. What really matters is the right term of (6.53) which is proportional to

$$\hat{f}_D(\mathbf{d}_0) = \frac{1}{M(M - 1)} \sum_{q=1}^{M(M-1)} e^{-\frac{1}{2}\|\mathbf{d}_0 - \mathbf{d}_q\|^2/(2h^2)^d} \tag{6.54}$$

where f_D denotes the unknown radially symmetric PDF of the $\mathbf{d}_{ij}$. Since we are interested in maximizing $\hat{V}_{R_2}$ we are consequently interested in having the highest possible value of $\hat{f}_D(\mathbf{d}_0)$. This will happen when vector $\mathbf{d}_0$ is aligned in such a way that its end tip falls in a highly peaked region of f_D. In this case $\mathbf{d}_0$ stands along the dominant structuring direction.

Example 6.3. Table 6.14 shows the entropic dissimilarity matrix for the 14 points of Fig. 6.21, using a 9-neighborhood. The points of Fig. 6.21 are referenced left to right and top to bottom as 1 to 14; point Q is, therefore, point 11.

Table 6.14 Entropic dissimilarity matrix for Fig. 6.21 example.

Points	1	2	3	4	5	6	7	8	9	10	11	12	13	14
1		8.24	8.37	8.82	9.61			8.56	8.61	8.81	9.17	10.21		
2	8.46		8.46	8.55	8.87			8.71	8.75	8.77	8.80	9.29		
3	8.62	8.54		8.54	8.62			8.90	8.80	8.80	8.71	8.94		
4		8.61	8.53		8.53	8.61			8.92	8.81	8.70	8.81	8.92	
5			8.62	8.54		8.54	8.62			8.94	8.71	8.80	8.80	8.90
6			8.87	8.55	8.46		8.46			9.29	8.80	8.77	8.75	8.71
7			9.61	8.82	8.37	8.24				10.21	9.17	8.81	8.61	8.56
8	8.56	8.59	8.77	9.27	10.16				8.24	8.36	8.84	9.59		
9	8.72	8.74	8.75	8.89	9.24			8.45		8.45	8.56	8.86		
10	8.93	8.81	8.79	8.80	8.90			8.61	8.52		8.53	8.61		
11		8.83	8.73	8.72	8.73	8.83			8.66	8.58		8.58	8.66	
12			8.90	8.80	8.79	8.81	8.93			8.61	8.53		8.52	8.61
13			9.24	8.89	8.75	8.74	8.72			8.86	8.56	8.45		8.45
14			10.16	9.27	8.77	8.59	8.56			9.59	8.84	8.36	8.24	

To compute, in this example, the dissimilarity values for, say, point 1, one starts by determining its 9 nearest neighbors and then compute the 9 entropies for each connection from point 1 to each nearest neighbor. For this reason, each row of the entropic dissimilarity matrix has only 9 values. Also, for the same reason, the entropic dissimilarity matrix is not symmetric since the 9 values are related to the nearest neighbors of each point and the nearest neighbors from 2 points a and b are not necessarily the same.   □

6.4.4.2 The LEGClust Proximity Matrix

The concept of dissimilarity matrix was formalized in 6.4.3. Using a dissimilarity matrix [4] one can build a proximity matrix, $\mathbf{L}$, where each ith row represents the dataset objects, each jth column the proximity order (1st column=closest object, ..., last column=farthest object), and each matrix

---

[4] $H_{R_2}(D, \mathbf{q}_i)$ does not verify all conditions of a common distance measure since it is not symmetric and may not verify the triangular inequality; it is, however, sufficient to build a proximity matrix.

element the object reference that, according to row object i, is in the jth proximity position.

Example 6.4. Let us consider a simple example of a dissimilarity matrix, actually an entropic one, presented in Table 6.15. This dissimilarity matrix is unconventional because each row contains only five dissimilarity values. The reason for this was given earlier.

Table 6.15 An example of a dissimilarity matrix for a simple dataset with 10 elements.

Points	1	2	3	4	5	6	7	8	9	10
1		3.08	3.63	3.38	3.19	4.69				
2	3.42		3.44	3.93	3.89		3.82			
3		3.30		3.49	3.66	3.63	3.31			
4		3.64	3.49			3.39	3.30	3.22		
5	3.38	3.22	3.98	3.31		3.49				
6			3.42	3.34			3.60	3.35	3.37	
7			3.12	3.28		3.43		3.20	3.41	
8			3.42			3.64	3.41		3.44	3.39
9				3.63		3.29	3.89	3.33		3.28
10			4.42			3.81	3.26	3.92	3.10	

The respective proximity matrix is presented in Table 6.16. Matrix elements of each row are related with each point in terms of dissimilarity. For instance, the first row ranks dissimilarities with respect to point 1: point 2 is the most similar to point 1, followed by points 5, 4, 3 and 6, respectively. L1 column elements correspond to points that have the smallest dissimilarity value w.r.t. each point referenced by the row number, followed by the points of columns L2 to L5 in decreasing order. □

Table 6.16 The entropic proximity matrix computed from Table 6.15.

Points	L1	L2	L3	L4	L5
1	2	5	4	3	6
2	1	3	7	5	4
3	2	7	4	6	5
4	7	6	5	3	2
5	2	4	1	6	3
6	4	8	9	3	7
7	3	8	4	9	6
8	10	7	3	9	6
9	10	6	8	4	7
10	9	7	6	8	3

Example 6.5. Table 6.17 shows the proximity values for point Q of Fig. 6.21. Rényi's quadratic entropy was computed as in (6.51).

One sees that the most similar point is point 10, i.e., point P in Fig. 6.21. In other words, vector PQ is the one that less changes the information potential of the 9-neighborhood vector field of Q. □

Table 6.17 Entropic proximities relative to point $Q(11)$ of Fig. 6.21.

Point	L1	L2	L3	L4	L5	L6	L7	L8	L9
11	10	12	9	13	4	3	5	6	2

One may consider that the proximity matrix defines connections between each point and the points referenced by each column: each point is connected to all other points following the order defined by the proximity matrix.

Each column of the proximity matrix corresponds to a layer of connections. A layer of connections can be represented as an unweighted subgraph, where each edge represents the connection between an object (connection set) and the corresponding object of that layer.

Example 6.6. We now present an example of a first layer unweighted subgraph based on a dissimilarity matrix and proximity matrix built with the Euclidian distance. Figure 6.24b represents the first layer subgraph for the dataset shown in Fig. 6.24a. Clusters formed with this first layer subgraph are called *elementary clusters.*

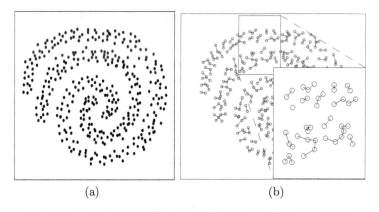

(a)   (b)

Fig. 6.24 Connections based on Euclidian distance for the double spiral dataset.

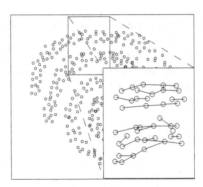

Fig. 6.25 "Ideal" connections for the spiral dataset of Fig. 6.24a.

As can be seen in Fig. 6.24b, the first layer connections take no account of the local structuring direction of the dataset. In Fig. 6.25 we present what one usually thinks should be the "ideal" connections, reflecting the local structuring directions of the data.

With classical distance measures, this behavior is not achieved. With an entropic dissimilarity measure the connections will follow the local structure of the dataset as we shall briefly see.   □

Example 6.7. Figure 6.26 shows the first layer connections for the dataset of Fig. 6.21, where one can see the difference between using a dissimilarity matrix based on Euclidian distance (Fig. 6.26a) or based on the LEGClust entropic measure (Fig. 6.26b). First layer connections, when using an entropic measure, clearly follow an horizontal line[5] and, despite the fact that point k_1 is the closest one, the stronger connection for point Q is the connection between $Q(11)$ and $P(10)$, as expected and shown in Table 6.17.

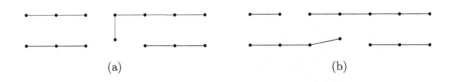

(a)   (b)

Fig. 6.26 First layer connections using a dissimilarity matrix based on Euclidian distance (a) and on entropic measure (b).

---

[5] The first connections shown in Fig. 6.26a are horizontal because in case of ties the algorithm chooses the first position of the tied points list. Points 2 and 8 are at the same distance to point 1 but the algorithm chose point 2 because it is the closest to point 1 in the list of points.

In Fig. 6.27 we present the LEGClust connections from layers L2 to L5. The connection from point Q to the closest element in terms of Euclidian distance only happens in layer L5 reinforcing the idea that the first LEGClust connections are affected by the local structure. We obtain the same behavior for the connections of all the layers, favoring the union of those clusters that follow the local structure of the data.   □

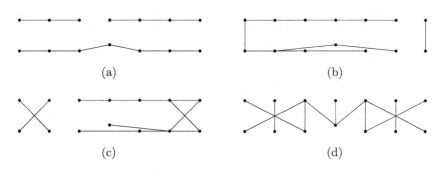

(a)   (b)

(c)   (d)

Fig. 6.27 The connections from 2nd (a), 3rd (b), 4th (c) and 5th (d) layers. Element $Q(11)$ connects to elements 12, 9, 13 and 4 respectively and according to values of Table 6.17.

Example 6.8. Figure 6.28 shows another example of the first layer subgraph obtained from an entropic proximity matrix. The connections clearly follow the local structure of the dataset.   □

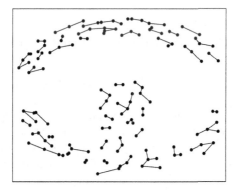

Fig. 6.28 Example of the first layer connections using an entropic dissimilarity measure.

6.4.4.3   The Clustering Process

With LEGClust dissimilarity matrix one can use any adequate algorithm to
cluster the data. LEGClust original algorithm described in [204] uses a hierarchical, agglomerative approach based on layered entropic (unweighted)
subgraphs built with the information given by the entropic proximity matrix (EPM). Examples of such subgraphs were already shown in Figs. 6.26b
and 6.27. The subgraph is built by connecting each object with the corresponding object of each layer (column) of the EPM. Using these subgraphs
one can hierarchically build the clusters by joining together the clusters that
correspond to the layer subgraphs with a predefined number of connections
between them.

Example 6.9. As an example to illustrate the clustering procedure we use the
simple two dimensional dataset presented in Fig. 6.29.

   This dataset consists of 15 points apparently constituting 2 clusters with
10 and 5 points each.

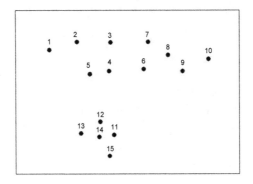

Fig. 6.29 A simple two dimensional dataset to illustrate the clustering procedure.

   Table 6.19 presents the EPM built from the entropic dissimilarity matrix
of Table 6.18.

   The EPM defines the connections between each point and those points in
each layer: point 1 is connected with point 2 in the first layer, with point 5 in
the second layer, with point 4 in the third layer and so on (see Table 6.19).

   The clustering process starts by defining the elementary clusters obtained
by connecting, with an oriented edge, each point with the corresponding point
of the first layer (Fig. 6.30a). There are 4 elementary clusters in our example.

   The second step of the algorithm connects, with an oriented edge, each
point with the corresponding point of the second layer (Fig. 6.30b). In order
to build the second step clusters we apply a rule based on the number of
connections to join each pair of clusters. We can use the simple rules of, a),

Table 6.18 The dissimilarity matrix for Fig. 6.29 dataset.

Points	1	2	3	4	5	6	7	8	9	10	11	12	13	14	15
1		4.08	4.63	4.48	4.19	5.69									
2	4.52		4.55	4.93	4.89		4.87								
3		4.40		4.59	4.66	4.64	4.41								
4		4.65	4.59		4.39	4.40	4.72								
5	4.48	4.72	4.98	4.41		4.59									
6			4.52	4.35			4.60	4.45	4.35						
7			4.17	4.78		4.53		4.20	4.51						
8			4.57			4.65	4.51		4.55	4.49					
9				4.63		4.29	4.89	4.43		4.28					
10		5.57				4.81	4.26	3.97	4.10						
11					4.84							4.23	4.53	4.24	4.28
12					4.78						4.34		4.43	4.33	4.59
13					4.84						4.48	4.26		4.17	4.40
14					4.88						4.38	4.36	4.40		4.38
15					5.33						4.05	4.33	4.33	4.02	

Table 6.19 The entropic proximity matrix for Fig. 6.29 dataset.

Points	L1	L2	L3	L4	L5
1	2	5	4	3	6
2	1	3	7	5	4
3	2	7	4	6	5
4	5	6	3	2	7
5	4	1	6	2	3
6	9	4	8	3	7
7	3	8	9	6	4
8	10	7	9	3	6
9	10	6	8	4	7
10	8	9	7	6	3
11	15	12	14	13	5
12	14	11	13	15	5
13	14	12	15	11	5
14	12	11	15	13	5
15	14	11	12	13	5

joining each cluster with the ones having at least k connections with it, or b), joining each cluster with the one having the highest number of connections with it, not less than a predefined k. In the experiments reported in [204], this second rule proved to be more reliable, and the resulting clusters were usually "better" than using the first rule. In our simple example we chose to join the clusters with the maximum number of connections not less than 2 ($k \geq 2$). In the second step 2 clusters are formed by joining clusters 1, 2 and 3 having at least 2 edges connecting them. Note that, apparently, there is only one connection between each pair of clusters but, as one can confirm in the EPM, these are double connections and therefore count as two.

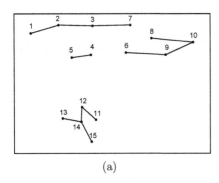 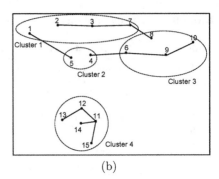

(a)   (b)

Fig. 6.30 First steps of a clustering process: a) first layer connections and resulting elementary clusters; b) second layer connections.

The process is repeated and the algorithm stops when only one cluster is present or when the same number of clusters in consecutive steps is obtained. The resulting number of clusters for this simple example was 4-2-2-2-1, with the same number of clusters (2) in steps 2, 3 and 4; therefore, one would normally consider 2 as the acceptable number of clusters.   □

Several refinements of the algorithm are also proposed in [198], namely strategies to avoid outliers, noise and micro clusters and different ways of choosing the number of connections to join clusters.

6.4.4.4 Parameter Selection

Number of Nearest Neighbors

The first parameter one must choose in the clustering process is the number of nearest neighbors (M). There is no specific rule for this choice. However, since the maximum number of steps in the clustering process is related to the number of nearest neighbors, one should not choose a very small value, because a minimum number of steps is needed to guarantee reaching a solution. Choosing a relatively high value for M is also not a good alternative because one loses information about the local structure, which is the main focus of the algorithm.

Based on the experiments reported in [204] in several datasets, a rule of thumb of using an M value not higher than 10% of the dataset size seems appropriate. Note that, since entropy computation has complexity $O\left(N\binom{M}{2}^2\right)$, the value of M has a large influence on the computational time. Hence, for large datasets a smaller M is recommended, down to 2% of the data size.

The Smoothing Parameter

The smoothing parameter h is very important when computing the entropy. In other works, [117, 84], using Renyi's quadratic entropy to perform clustering, it is assumed that the smoothing parameter is experimentally selected and that it must be fine-tuned to achieve acceptable results. Formula (6.8), $h_{fat} = 25\sqrt{c/n}$, was proposed in [203] and showed to produce good results in neural network classification using error entropy minimization, as mentioned in Sect. 6.1.1.1. For the LEGClust algorithm we need a formula that reflects the standard deviation of the data. Following the approach described in 6.1.1.1, a new formula, inspired on (6.7), was proposed in [198]:

$$ h_{op} = 2\,\overline{s}\left(\frac{4}{(d+2)n}\right)^{\frac{1}{d+4}}, \tag{6.55} $$

where $\overline{s}$ is the mean value of the sample standard deviations for all d dimensions. All experiments with the entropic clustering algorithm reported in [204] were performed using formula (6.55).

Although the value of the smoothing parameter is important, it is not crucial to obtain good results. As we increase the h value, the kernel becomes smoother and the entropic proximity matrix becomes similar to the Euclidian distance proximity matrix. Extremely small values of h will produce undesirable behaviors because the entropy will have high variability. Using h values in a small interval, near the h_{fat} value, does not affect the final clustering results.

Minimum Number of Connections

The minimum number of connections, k, to join clusters in consecutive steps of the algorithm is the third parameter that must be chosen. One should not use $k = 1$ to avoid outliers and noise, especially if they are located between clusters. If the elementary clusters have a small number of points, high values for k are also not recommended because the impossibility of joining clusters could then arise due to lack of a sufficient number of connections. Experimental evidence provided in [204] shows that good results are obtained when using either $k = 2$ or $k = 3$.

An alternative is simply to join at each step the two clusters with the highest number of connections between them.

6.4.5 Experiments

The LEGClust algorithm was applied to a large variety of artificial and real-world datasets, some of them with a large number of features. We now present the results of some of these experiments, which are described in [204] and involve the real-world datasets summarized in Table 6.20. Dataset NCI Microarray can be found in [163], 20NewsGroups, Dutch Handwritten Numerals (DHN), Iris, Wdbc and Wine in [13] and Olive in [75].

Table 6.20 Real datasets used in the experiments.

Dataset	# Objects	# Features	# Classes
20NewsGroups	1000	565	20
DHN	2000	3	10
Iris	150	4	3
NCI Microarray	64	6830	12
Olive	572	8	9
Wdbc	569	30	2
Wine	178	13	3

The artificial two-dimensional datasets are taken from [201] and can be found in [197]. These datasets are used to better visualize and control the clustering process. Some examples are depicted in Fig. 6.31.

For the artificial dataset problems the clustering solutions yielded by different algorithms were compared with the majority choice solutions obtained in the human clustering experiments mentioned in Sect. 6.4.4 and described in [201]. For the real-world datasets the comparison was made with the supervised classes. In both cases — majority choice or supervised classes — we will refer to these solutions as reference solutions or reference clusters.

The LEGClust solutions were compared with those of the following well-known clustering algorithms: Chameleon algorithm, included in the software package Cluto [126], two Spectral clustering algorithms (Spectral-Ng [166] and Spectral-Shi [209]) and one density-based algorithm, DBScan [248].

Regarding the experiments with the artificial datasets, shown in Fig. 6.31, Fig. 6.32 presents the solutions obtained with LEGClust.

In Fig. 6.33 we present the solutions obtained with the Chameleon algorithm that differ from those suggested by LEGClust.

From the performed experiments, an important aspect noticed when using the Chameleon algorithm was the different solutions obtained for slightly different parameter values. Dataset 6.33c was reported in [204] as being the one that presented more difficulties in tuning the parameters involved in Chameleon algorithm. Such tuning difficulties don't arise when using LEGClust, since as we said before LEGClust is not sensitive to small changes of its parameters. A particular difference between the Chameleon and the LEGClust corresponds to the curious solution given by Chameleon Fig. 6.33b.

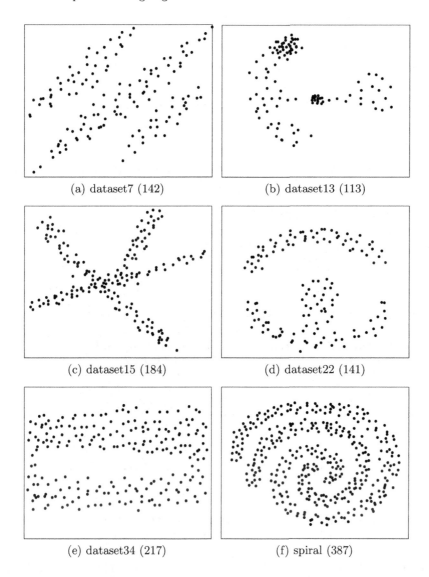

(a) dataset7 (142)   (b) dataset13 (113)

(c) dataset15 (184)   (d) dataset22 (141)

(e) dataset34 (217)   (f) spiral (387)

Fig. 6.31 Some artificial datasets used in the experiments, reported in [201] (the number of points is given in parentheses).

When constraining the number of clusters to 3, this solution is the only solution not suggested by humans that performed the tests mentioned in Sect. 6.4.4. The solutions for this same problem given by LEGClust algorithm are shown in Figs. 6.32a and 6.32b.

The spectral clustering algorithms gave some good results for some datasets, but were unable to resolve some non-convex datasets like the double spiral problem (Fig. 6.34 and 6.35), which LEGClust did.

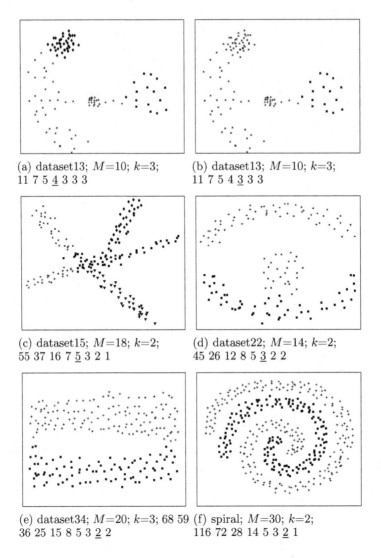

(a) dataset13; $M{=}10$; $k{=}3$;
11 7 5 <u>4</u> 3 3 3

(b) dataset13; $M{=}10$; $k{=}3$;
11 7 5 4 <u>3</u> 3 3

(c) dataset15; $M{=}18$; $k{=}2$;
55 37 16 7 <u>5</u> 3 2 1

(d) dataset22; $M{=}14$; $k{=}2$;
45 26 12 8 5 <u>3</u> 2 2

(e) dataset34; $M{=}20$; $k{=}3$; 68 59
36 25 15 8 5 3 <u>2</u> 2

(f) spiral; $M{=}30$; $k{=}2$;
116 72 28 14 5 3 <u>2</u> 1

Fig. 6.32 The clustering solutions for each dataset suggested by LEGClust. Each label shows: the dataset name; the number of neighbors (M); the number of connections to join clusters (k); the number of clusters found at each step of the algorithm (underlined is the step of the shown solution).

The DBScan algorithm failed in finding the reference clusters in all artificial datasets (except the one of Fig. 6.31a).

The results reported in [204] for all algorithms applied to the artificial datasets lead to the conclusion that, as expected, the solutions obtained with the density-based algorithms are worse than those obtained with any of the other algorithms. The best results were obtained with LEGClust and Chameleon.

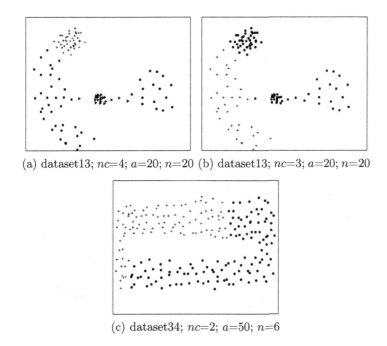

(a) dataset13; $nc=4$; $a=20$; $n=20$ (b) dataset13; $nc=3$; $a=20$; $n=20$

(c) dataset34; $nc=2$; $a=50$; $n=6$

Fig. 6.33 Some clustering solutions suggested by Chameleon. The considered values nc, a and n are shown in each label.

We now report on the experiments with real-world datasets described in [204].

The DHN dataset consists of 2000 images of handwritten numerals ('0'–'9') extracted from a collection of Dutch utility maps [60]. A sample of this dataset is depicted in Fig. 6.36. In this dataset, the first two features represent the pixel position and the third one, the gray level. Experiments with this dataset were performed with LEGClust and Spectral clustering.

Results are presented in Table 6.21. ARI stands for Adjusted Rand Index, a measure for comparing results of different clustering solutions when the labels are known [107]. This index is an improvement of the Rand Index [180], it lies between 0 and 1 and the higher the ARI index the better the clustering solution. The parameters for both algorithms were tuned to give the best possible solutions. In this problem, LEGClust performs far better than Spectral-Shi and with similar (but slightly better) results than Spectral-Ng. Table 6.21, shows different LEGClust results for different choices of the minimum number of connections (k) to join clusters. These results clearly show that different values of k produce results with small differences in the ARI value.

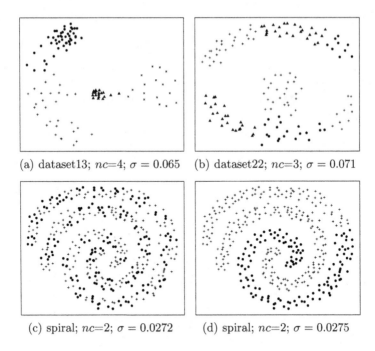

(a) dataset13; nc=4; $\sigma = 0.065$ (b) dataset22; nc=3; $\sigma = 0.071$

(c) spiral; nc=2; $\sigma = 0.0272$ (d) spiral; nc=2; $\sigma = 0.0275$

Fig. 6.34 Some clustering solutions given by Spectral-Ng. Each label shows: the dataset name; the pre-established number of clusters and the σ value.

Table 6.21 Results and parameters used in the comparison of LEGClust and Spectral clustering in experiments with DHN, 20NewsGroups and NCI Microarray datasets.

	LEGClust			Spectral-Ng			Spectral-Shi		
	M	k	ARI	nc	σ	ARI	nc	σ	ARI
DHN	30	10	0.628	10	10	0.287	10	12	0.573
	30	8	0.608						
	30	12	0.574						
20NewsGroups	20	3	0.289	20	12	0.479	20	20	0.006
	20	2	0.287						
NCI Microarray	4	2	0.148	3	80	0.177	3	10	0.138
	6	3	0.148						
	10	3	0.148						

In Table 6.22 we show an example of a confusion matrix obtained with LEGClust for an experiment with the DHN dataset.

In the experiments with the 20NewsGroups dataset, a random sub-sample of 1000 elements from the original dataset was used. This dataset is a 20 class text classification set obtained from 20 different news groups. The dataset was prepared by stemming words according to the algorithm described in

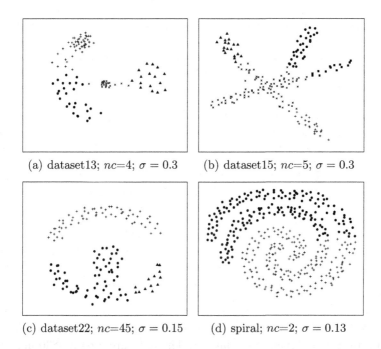

(a) dataset13; $nc=4$; $\sigma = 0.3$ (b) dataset15; $nc=5$; $\sigma = 0.3$

(c) dataset22; $nc=45$; $\sigma = 0.15$ (d) spiral; $nc=2$; $\sigma = 0.13$

Fig. 6.35 Some clustering solutions given by Spectral-Shi. Each label shows: the dataset name; the pre-established number of clusters and the σ value.

Fig. 6.36 A sample of the Dutch Handwritten Numerals dataset.

[173]. The size of the corpus (the number of different words present in all the stemmed dataset) defines the number of features. In this sub-sample, only the words that occur at least 40 times were considered, thus obtaining a corpus of 565 words. The results of the experiments with LEGClust and Spectral clustering are shown in Table 6.21.

Table 6.22 A confusion matrix of a clustering solution obtained with LEGClust for the DHN dataset.

				Classes						
	176	169	3	0	0	1	3	0	1	9
	1	19	0	0	0	0	0	0	0	0
	7	1	188	1	16	2	178	14	3	6
	0	3	3	195	83	2	0	0	5	0
Clusters	1	0	3	4	101	10	4	1	7	0
	0	2	2	0	0	167	0	3	1	1
	0	1	0	0	0	2	0	86	0	1
	0	2	1	0	0	8	0	95	3	1
	0	0	0	0	0	8	14	0	180	1
	15	3	0	0	0	0	1	1	0	181

The NCI Microarray dataset is a human tumor microarray data and an example of a high-dimensional dataset. The data are a 64×6830 matrix of real numbers, each representing an expression measurement for a gene (column) and a sample (row). There are 12 different tumor types, one with just 1 representative and three with 2 representatives. Experiments were performed to compare the results from LEGClust with those obtained by Spectral clustering. The final number of clusters for both algorithms was chosen to be 3, following the example described in [94]. Results are shown in Table 6.21.

Table 6.23 Results and parameters of LEGClust and Chameleon in experiments on 4 real-world datasets.

	Chameleon			LEGClust		
	a	n	ARI	M	k	ARI
Iris	9	50	0.658	15	3	0.750
Olive	40	40	0.733	25	3	0.616
Wdbc	40	25	0.410	20	3	0.574
Wine	30	21	0.400	15	3	0.802

Results presented in Table 6.21 show that LEGClust performs better than Spectral-Shi algorithm in the three datasets and, compared with Spectral-Ng, it achieves better results in the DHN dataset and similar ones in the NCI Microarray.

Table 6.23 shows the clustering solutions obtained by LEGClust and by Chameleon in the experiments with the datasets Iris, Olive, Wdbc and Wine. The parameters used for each experiment are also shown in Table 6.23. The final number of clusters is the same as the number of classes. Results with LEGClust are better than those obtained with Chameleon, except for dataset Olive.

6.5 Task Decomposition and Modular Neural Networks

This section addresses a complex type of classifiers: the modular neural network (MNN) using task decomposition performed by a clustering algorithm. We will specifically present the algorithmic description and results of MNNs built with MEE MLPs (with Rényi's quadratic entropy as risk functional) and with the LEGClust algorithm presented in Sect. 6.4 performing the task decomposition [200].

Task decomposition is one of the strategies used to simplify the learning process of any learning system. It basically consists of partitioning the input space into several regions, this way decomposing the initial problem into different subproblems. This is done based on the assumption that these regions possess different characteristics and so they should be learned by specialized classifiers. By subsequent integration of the learning results, one is hopefully able to achieve better solutions for the initial problem. Generally, task decomposition can be obtained in three different ways: explicit decomposition (the task is decomposed by the designer before training), class decomposition (the decomposition is made based on the classes of the problem) and automatic decomposition. When automatic decomposition is used, it can either be made during the learning stage or before training the modules using a clustering algorithm.

6.5.1 Modular Neural Networks

A modular neural network (MNN) is an ensemble of learning machines. The idea behind this kind of learning structure is the divide-and-conquer paradigm: the problem should be divided into smaller sub-problems that are solved by experts (modules) and their partial solutions should be integrated to produce a final solution (Fig. 6.37).

Ensembles of learning machines were often proved to give better results than single learners. The proofs are mainly empirical [21, 54] but there are some theoretical results [129, 8, 5] that support this assumption if some conditions are satisfied.

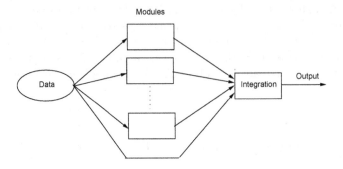

Fig. 6.37 A modular neural network

To use a MNN, three stages have to be considered:

- The **task decomposition**, where the problem is divided into smaller problems, each one to be presented to one of the modules or expert networks. To better understand the task decomposition process let us take a look at the artificial dataset shown in Fig. 6.38. This is a three-class problem where the input space is clearly divided into two regions: one of the regions (upper right) contains samples from two classes (crosses and circles) and the other contains samples from all three classes. Note that there are two classes with samples belonging to the two different regions. By having a classifier dedicated to each region we are able to transform this particular problem into two simpler ones.

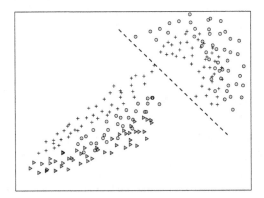

Fig. 6.38 The partition of the input space for a three-class problem.

- The *training phase*, where each individual expert (module) is trained until it learns to solve its particular sub-problem.
- The *decision integration*. Corresponds to combining the outputs of the experts, producing an integrated network output. There are three approaches one can use to accomplish this task: gating network [110]; module votes [14]; hierarchical integration (which can also use voting and/or gating networks) [123,111]. The above cited work [200] uses a gating network, which works as an additional expert trained to recognize the region of the input space where each of the experts have their regions of expertise, defined in the task decomposition phase.

Once the learning process has finished and a new, unclassified, pattern is presented to the network, the individual experts compute the class it might belong, but the gate network selects only a particular output from the expert it considers to be 'competent' to solve the problem, taking into account the region of the input space to which the pattern belongs.

As mentioned earlier, task decomposition learning is done before training the modules. This is usually accomplished by a clustering algorithm [231,5,4, 64]. There are several well-known algorithms to perform clustering, being the most common ones those based on matrix theory and graph theory. However, as mentioned in Sect. 6.4, it is also known that these and other kinds of algorithms often have serious difficulties in identifying clusters that reflect the local structure of the data objects. In Sect. 6.4 we showed how the LEGClust algorithm successfully partitions the data into meaningful clusters, reflecting the local structuring of the data. These are the basic reasons why the MNN classifier described in [200] uses the LEGClust algorithm — a MEE-flavored clustering approach — for the task decomposition. Moreover, it uses MEE MLPs as expert modules.

6.5.2 Experiments

A considerable number of classification experiments with several datasets were performed and reported in [200], using modular neural networks with task decomposition carried out by three different clustering algorithms: k-means (K-MNN), Spectral Clustering (S-MNN) [166], and LEGClust (EC-MNN). All neural networks used in the experiments, both the modules and the gates of the MNN, were MLPs with one hidden-layer, trained with the R_2EE risk functional. The MLPs were trained with the backpropagation algorithm and early stopping. The experimental validation was made with the holdout method: two runs with half the dataset used for training and the other half for testing, swapping the roles in the second run. Each module was trained with the input data selected by LEGClust (each module learns the data from each cluster). The gate network was trained with all the data labeled by LEGClust.

6.5.2.1 Datasets

The experiments reported in [200] used several real-world datasets and also the artificial one depicted in Fig. 6.38 called ArtificialF2. The real-world datasets are all publicly available: the Breast Tissue, CTG, Diabetes and Sonar datasets can be found in [13], PB12 in [110] and Olive in [75]. Table 6.24 contains a summary of the characteristics of these datasets showing the number of data instances, number of features and the number of classes for each dataset.

Table 6.24 The datasets used in the experiments reported in [200].

Dataset	# samples	# features	# classes
ArtificialF2	222	2	3
Breast Tissue	106	9	6
CTG	2126	22	10
Diabetes	768	8	2
Olive	572	8	9
PB12	608	2	2
Sonar	208	60	2

6.5.2.2 Results

The results we present are reported in detail in [200]. Table 6.25 presents the parameters of each modular neural network for the results shown in Table 6.26. For each type of MNN Table 6.25 shows the number of experts and, for each of them, the number of hidden neurons and of output neurons (the number of classes in each cluster). The MLP topologies are denoted $[d : n_h : c]$, as in Sect. 6.1.1, where d is the number of inputs, n_h the number of hidden neurons, and c the number of classes.

The presented topologies correspond to the best results of a large number of experiments with different combinations of the number of neurons in each module and in the gate.

The results in Tables 6.26 and 6.27 are averages and standard deviations of the test set error rates for 20 repetitions of each experiment [200]. Table 6.26 shows in bold the result for the winning algorithm; the one achieving the lowest average error rate. For these datasets EC-MNN is the algorithm that wins more often. Moreover, EC-MNN outperforms K-MNN for the dataset ArtificialF2, Breast Tissue, CTG, and Olive in a statistically significant way (with t-test p value of 0.037 for ArtificialF2 and ≈ 0 for the other ones), whereas K-MNN outperforms EC-MNN in only one case (Sonar with t-test $p = 0.018$).

Table 6.25 Topologies of the modular neural networks.

Dataset	# experts	K-MNN
ArtificialF2	3	[2:18:3] [2:18:2] [2:18:2]
Breast Tissue	3	[9:10:2] [9:12:2] [9:12:2]
CTG	4	[22:22:10] [22:18:9] [22:18:9] [22:26:3]
Diabetes	4	[8:10:2] [8:12:2] [8:10:2] [8:12:3]
Olive	4	[8:6:6] [8:12:4] [8:12:8] [8:12:3]
PB12	3	[2:6:2] [2:6:2] [2:2:2]
Sonar	3	[60:12:2] [60:12:2] [60:14:2]

Dataset	# experts	S-MNN
ArtificialF2	3	[2:18:3][2:18:2][2:16:2]
Breast Tissue	3	[9:4:2][9:14:2][9:6:2]
CTG	4	[22:18:10][22:22:10][22:22:10][22:22:3]
Diabetes	3	[8:18:2][8:12:2][8:10:3]
Olive	4	[8:12:4][8:12:4][8:4:3][8:6:3]
PB12	3	[2:5:2][2:5:2][2:2:2]
Sonar	3	[60:10:2][60:16:2][60:16:2]

Dataset	# experts	EC-MNN
ArtificialF2	3	[2:20:3][2:12:2][2:14:2]
Breast Tissue	3	[9:12:2][9:12:2][9:3:2]
CTG	4	[22:20:6][22:18:2][22:26:2][22:26:3]
Diabetes	4	[8:14:2][8:18:2][8:12:2][8:16:3]
Olive	4	[8:10:4][8:4:3][8:12:2][8:8:3]
PB12	3	[2:4:2][2:5:2][2:2:2]
Sonar	3	[60:12:2][60:12:2][60:12:2]

Table 6.26 Average error rates (with standard deviations) for MNNs using different clustering algorithms. Best results (lowest error rate) in bold.

Dataset	K-MNN	S-MNN	EC-MNN
ArtificialF2	16.40 (2.40)	15.32 (3.55)	**14.70** (3.22)
Breast Tissue	58.95 (7.54)	33.53 (4.47)	**32.79** (3.72)
CTG	22.90 (0.86)	23.91 (2.91)	**20.67** (2.38)
Diabetes	24.45 (1.45)	23.96 (1.76)	**23.89** (1.64)
Olive	49.11 (2.89)	5.20 (1.11)	**4.74** (0.89)
PB12	**7.23** (1.17)	7.28 (0.95)	7.25 (0.80)
Sonar	**16.14** (3.43)	23.69 (4.57)	18.57 (3.40)

Table 6.27 shows the errors obtained using single neural networks (SNNs — MLP's with one hidden layer). For the studied datasets the SNN solution outperformed the MNN solution in only one dataset (CTG). In three other datasets (ArtificialF2, Olive and Sonar) the SNN performance was statistically significantly worse that the best MNN solution (t-test p values below 0.005). We remind that the MNN approach is only effective, achieving better

Table 6.27 Average error rates (with standard deviations) for the experiments with single neural networks (SNN).

Dataset	SNN	n_h
ArtificialF2	19.56 (3.95)	20
Breast Tissue	32.75 (3.26)	22
CTG	15.70 (0.60)	20
Diabetes	23.90 (1.69)	15
Olive	5.45 (0.62)	15
PB12	7.51 (0.37)	6
Sonar	21.90 (2.80)	14

results than single neural networks, when the $X \times T$ space possesses some divisive properties.

6.6 Decision Trees

Decision trees (also known as classification trees) are attractive for applications requiring the semantic interpretation assignable to nodal decision rules, for instance as diagnostic tools in the medical area. We will only be interested in binary decision trees based on univariate splits, by far the most popular type of decision trees. These classifiers have a hierarchical structure such that any data instance x travels down the tree according to the fulfillment of a sequence of binary (dichotomous) decisions (does x belong to ω_k or to $\overline{\omega}_k$?), evaluated on the basis of single variables. The traveling down stops at a tree leaf where the respective class label is assigned to x.

Tree construction algorithms typically follow a greedy approach: find at each node the best univariate decision rule, according to some criterion. Denoting the j-th tree node by u_j, a univariate split represents a binary test z_j as $\{x_{ij} < \Delta_j, z_j(x_i) = \omega_k; \overline{\omega}_k$ otherwise$\}$ for *numerical* inputs (i.e., real-valued inputs) or as $\{x_{ij} \in B_j, z_j(x_i) = \omega_k; \overline{\omega}_k$ otherwise$\}$ for *categorical* inputs; Δ_j and B_j are, respectively, a numerical threshold and a set of categories. The search for the best univariate decision rule is, therefore, the search for the best triple $\{x_i, \omega_k, \Delta_j$ or $B_j\}$ for all possible combinations of features, classes, and feature values, at u_j; equivalently, the search for the *best data split*, since the training dataset at u_j is split into two training datasets, one sent to the left child node, u_{jl} (collecting, say, all data instances satisfying the split rule), and the other sent to the right node, u_{jr} (collecting the remaining instances). During tree construction (tree growing) the splitting of u_j into its children nodes u_{jl} and u_{jr} goes on until some stopping rule is satisfied. Typically node splitting stops when there is an insufficient number of instances. Literature on binary decision trees is abundant (see e.g., [33, 52, 80, 194, 188, 147, 161]).

Tree design approaches are aimed at finding "optimal" solutions: minimum sized trees with high classification accuracy [12, 130, 177]. The greedy approach is followed because a search on the whole set of possible trees for a given problem is almost always impractical (finding optimal binary trees is NP-complete, [108]). There is, however, a shortcoming with the greedy designs [130]; whereas node performance is a linear function of the class recognition rates, the overall tree performance is usually highly nonlinear. Since the greedy design is not guaranteed to be "optimal" in any sense, the accuracy, efficiency and generalization capability of the final solution relies on using some sort of *tree pruning* — i.e., removing subtrees with no statistically significant contribution to the final performance —, remedying the usual over-fitting of the final solution. Even with pruning, final tree solutions are often found to be more complex than they should [116].

The MEE trees we are about to describe are constructed relying on the (discrete) Shannon MEE data splitter, studied in Chap. 4, for the search of best splits. Compared to classic data splitting rules (presented in Sect. 4.1.4) the MEE approach has an important advantage: one is able to devise a stopping rule based on an implict distribution overlapping measure, such as the interval-end hit rate (Sect. 4.1.3.2). In other words, by using MEE splitters one is implicitly contributing to a better generalization, making the tree design less dependent on pruning.

6.6.1 The MEE Tree Algorithm

The greedy algorithm for growing a MEE tree has the following main steps [152]:

1. Starting at the root node (where the whole training set is used), at each tree node one has an $n \times d$ (n cases, d features) matrix X and an $n \times c$ (n cases, c classes) matrix T, filled with zeros and ones. A univariate split z (with parameter Δ or B) minimizing EE is searched for in the $d \times c$-dimensional space.

2. For that purpose, the error rates $\hat{P}_{10} = n_{10}/n$, $\hat{P}_{01} = n_{01}/n$ ($n_{tt'}$: number of class t cases classified as t') are computed for each candidate class label $t \in \{0, 1\}$. ($T = \{0, 1\}$ is computationally easier to use.)

3. The rule at node u_j — therefore, the triple $\{x_{.j}, t_{.k}, \Delta_j$ or $B_j\}$ — corresponding to the smallest empirical error entropy is assigned to the node and if a stopping criterion is satisfied the node becomes a leaf. Otherwise, the left and right node (u_{jl}, u_{jr}) sets are generated and steps 1 and 2 iterated.

The error rates at step 2 are computed for distinct values of every feature $x_{.j}$, $j = 1, \ldots, d$ serving as candidate split points, Δ_j for numerical-type rules (often the middle points of the original training set values are used), and B_j for combinations of categorical values in the case of categorical variables. A tree leaf is reached whenever a lower bound on the number of instances is reached or an interval-end hit occurs, signaling a large distribution overlap (Sect. 4.1.3.2). Instead of interval-end hits, other distribution overlapping criteria could be envisaged; for instance, one could attempt to detect a concave shape of EE (exemplified by Fig. 4.8b) as done in [151]. Final class label assignment of the leaves is made by the usual process of majority voting. The algorithm has also to take care of the following three issues specific of these trees.

6.6.1.1   Position of the Candidate Class

For numerical features we have seen in Sect. 4.1.4.2 that EE depends on the position of the class (i.e., on whether we decide ω_k for u_{jl} or u_{jr}). Thus, error rates must be computed for two class configurations: the candidate class ($t = 1$) corresponds to the $x_{ij} > \Delta_j$ rule; the candidate class ($t = 1$) corresponds to the $x_{ij} \leq \Delta_j$ rule. Note that the candidate class position achieving MEE is not known *a priori*.

It is, however, a trivial task to compute EE for one of the configurations, when the error rates have been computed for the other. Let us assume that the computation for the $x > \Delta$ configuration has been carried out for a nodal set with n instances. In pseudo code: $rule \leftarrow (x > delta)$; $n_{10} \leftarrow sum$ (not $rule$ and t) ; $n_{01} \leftarrow sum($ $rule$ and not $t)$. Using these n's one then computes their values for the other configuration simply as: $n_1 \leftarrow sum(t); n_{10} \leftarrow n_1 - n_{10}; n_{01} \leftarrow n - n_1 - n_{01}$.

The class position issue does not apply to categorical variables.

6.6.1.2   Handling Categorical Variables

MEE trees handle subsets of categorical features, of a set $B = \{b_1, \cdots, b_m\}$, in a simple way: the full computation of the MEE split point has only to be carried out for the m singleton categories (instead of having to perform 2^m computations). This represents a considerable time saving. We now prove this. Denoting a singleton category set $\{b\}$ by b and its complement, $B - \{b\}$, by $\bar{b}$, let us assume that we have computed

$$P_{10} = P(1, \bar{b}), \quad P_{01} = P(0, b) , \tag{6.56}$$

where $P(1, \bar{b})$ and $P(0, b)$ are short notation for $P(\omega_1, x \notin \{b\})$ and $P(\omega_0, x \in \{b\})$, respectively (with $\omega_0 \equiv \bar{\omega}_1$). The decision rule here is $r \doteq x \in \{b\}$.

We have already seen in formula (4.46) that in order to compute EE we only need the P_{10} and P_{01} values. These can also be used with class priors $p = P(\omega_1)$ and $q = 1 - p$ to compute:

$$P(\bar{b}|1) = P_{10}/p; \quad P(b|1) = 1 - P_{10}/p; \quad P(b|0) = P_{01}/q . \tag{6.57}$$

Let us now consider non-singleton sets, B_i. We first notice that:

$$P(\{B_1, B_2\}|\omega)) = P(B_1|\omega) + P(B_2|\omega) \text{ whenever } B_1 \cap B_2 = \varnothing . \quad (6.58)$$

Therefore, for any non-singleton B_i, we may write:

$$P(B_i|1) = \sum_{b \in B_i} P(b|1); \quad P(B_i|0) = \sum_{b \in B_i} P(b|0) . \quad (6.59)$$

Thus, once P_{10} and P_{01} have been computed for all m singleton sets, it is an easy task to compute them for any non-singleton category set B_i, using formulas (6.57) and (6.59), since for both theoretical and empirical probability mass functions pertaining to B_i:

$$P_{10} = p(1 - P(B_i|1)); \quad P_{01} = qP(B_i|0) . \quad (6.60)$$

6.6.1.3 Error Entropy of Class Unions

Since the MEE approach is a two-class discrimination approach, one expects to obtain performance improvements for datasets with $c > 3$ classes by considering class unions, i.e., by including merged classes, say of k classes with k up to $\lfloor c/2 \rfloor$ in the set of candidate classes. The difficulty thereof is that the number of candidate classes may become quite high. There is, however, a fast way of computing the dichotomous decision errors for unions of classes as we shall now show.

Consider the class union $\omega = \omega_1 \cup \omega_2$, $\omega_1 \cap \omega_2 = \varnothing$, and suppose that we have computed the following three quantities for ω_i, $(i = 1, 2)$ and for a decision rule r:

1. $n_{10}(\omega_i, \bar{r})$ — number of instances of class i that do not satisfy the rule;
2. $n_{11}(\omega_i, r)$ — number of instances of class i that satisfy the rule;
3. n_r — number of instances (from whatever class) that satisfy the rule.

$P_{10}(\omega_i)$ (i.e., P_{10} for the ω_i vs $\bar{\omega}_i$ decision) is, as before, simply $n_{10}(\omega_i, \bar{r})/n$. Let us now see how one can compute $P_{01}(\omega_i)$ using $n_{11}(\omega_i, r)$ instead of $n_{01}(\bar{\omega}_i, r)$.

First, notice that:

$$P(r) = P(\bar{\omega}_i, r) + P(\omega_i, r) ; \quad (6.61)$$

$$P(\omega_i, r) = P(\omega_i)P(r|\omega_i) = \frac{n_{11}(\omega_i, r)}{n} , \quad (6.62)$$

where n is the total number of instances at the node. From formulas (6.61) and (6.62) we derive:

$$P_{01}(\omega_i) = P(r) - P(\omega_i, r) = \frac{n_r}{n} - \frac{n_{11}(\omega_i, r)}{n} . \qquad (6.63)$$

Let us consider the class union ω. For both the theoretical and empirical PMF's the following holds:

$$P_{10}(\omega) = P(\omega, \bar{r}) = P(\omega_1, \bar{r}) + P(\omega_2, \bar{r}) = P_{10}(\omega_1) + P_{10}(\omega_2) ; \qquad (6.64)$$

$$P_{01}(\omega) = P(\bar{\omega}, r) = P(r) - P(\omega, r) = P(r) - [P(\omega_1, r) + P(\omega_2, r)] . \quad (6.65)$$

We now see why we need to compute $n_{11}(\omega_i, r)$ instead of $n_{01}(\bar{\omega}_i, r)$. It is simply a consequence of formula (6.65). We wouldn't be able to come out with a similar expression in terms of $n_{01}(\bar{\omega}_i, r)$. We also see that it is a negligible time-consuming task to compute the probabilities of interest for the union of classes, because n_r is computed only once for each feature and is independent of the candidate class, and all that remains to be done is the two additions and one subtraction of formulas (6.64) and (6.65).

6.6.2 *Application to Real-World Datasets*

The performance of MEE trees applied to real-world datasets was analyzed in [152, 153]. In both works the results of MEE trees were compared against those obtained with classic tree algorithms: the CART algorithm [33] using the classic splitting rules discussed in Sect. 4.1.4 (CART-GI, CART-IG, and CART-TWO, respectively for the Gini index, the information gain, and the Twoing criterion); the C4.5 algorithm [177]. These algorithms are still the ones in current extensive use and in available software tools of decision tree design.

All algorithms were run with unit misclassification costs (i.e., the error rates are weighted equally for all classes), estimated priors and the same minimum number of instances for a node to be split: 5. The CART and MEE algorithms were run with Cost-Complexity Pruning (CCP) with the 'min' criterion and 10-fold cross-validation [33]. The C4.5 algorithm was run with Pessimistic Error Pruning (PEP) at 25% confidence level [68]. References [68, 34] report a good performance of CCP-min over other pruning methods, including PEP, which has a tendency to underprune. CCP was also found to appropriately limit tree growth more frequently than other pruning methods [116].

Figure 6.39 shows the CCP-pruned MEE tree solution for the Glass dataset [13]; tree construction involved the consideration of class unions up to three classes (the Glass dataset has six classes).

The results reported in [152] respect to 36 public real-world datasets, quite diverse in terms of number of instances, features and classes as well as of

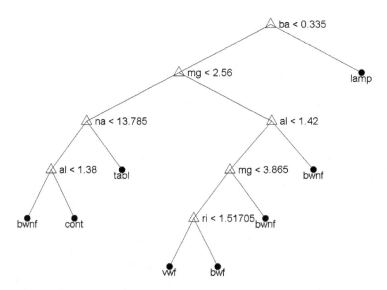

Fig. 6.39 CCP-pruned MEE tree for the Glass dataset. The decision rules are expressed in terms of inequalities involving data features (chemical elements and refractive index). A class label is shown below each tree leaf.

feature types. In this work resubstitution and leave-one-out estimates of the error were computed (except for a few large datasets where 5-fold stratified cross-validation was used).

In the following work [153] the study involved 42 datasets (from [13], except the colon, central nervous system and leukemia datasets which are from the Kent Ridge Biomedical Dataset [138]). All datasets were analyzed with 10-fold stratified cross-validation. In both works statistics regarding the tree sizes in the cross-validation experiments were also computed. The statistical methods used in the multiple comparison assessment of the algorithms followed recommendations in [51, 79, 249, 195] (see details in [152, 153]).

In the cited works [152, 153] no statistically significant difference among the algorithms was found in what regards error performance; however, a significant difference did emerge when comparing tree sizes. We now provide the main evidence on these issues.

Table 6.28 shows the cross-validation estimates of the error rate for the 42 datasets analyzed in [153], with the best MEE solution for class unions up to $\lfloor c/2 \rfloor$. The total number of wins (smallest error) and losses (highest error) are also shown in Table 6.28 with the chi-square test p: no significant difference is found relative to the equal distribution hypothesis. The Friedman test did not detect significant differences ($p = 0.453$). Other statistical tests confirmed these findings.

Table 6.29 shows the averages and ranges of tree sizes achieved in the cross-validation experiments by all algorithms. The total number of wins (smallest

Table 6.28 CV10 mean of test set P_e with wins (bold) and losses (italic). Standard deviations are reported in [153].

	Arrythmya	Balance	Car	Clev. HD2	Clev.HD5	CNS
Gini	0.3518	**0.1952**	0.0434	0.2357	*0.4680*	0.3500
Info Gain	0.3606	0.2528	0.0457	0.2593	**0.4512**	**0.3000**
Twoing	0.3628	0.2176	**0.0405**	*0.2761*	0.4613	0.3167
C4.5	*0.3934*	0.2192	0.0434	**0.1987**	0.4577	*0.4833*
MEE	**0.3208**	*0.3120*	0.0718	0.2222	0.4646	0.3667
	Colon	Cork stop.	Credit	CTG	Dermatol.	E-coli
Gini	*0.2581*	**0.1133**	0.1363	0.1689	**0.0531**	*0.1927*
Info Gain	*0.2581*	*0.1400*	0.1363	0.1877	0.0587	0.1896
Twoing	0.2419	0.1267	0.1363	0.1811	0.0670	0.1682
C4.5	0.2419	**0.1133**	**0.1332**	**0.1731**	*0.0991*	0.1713
MEE	**0.1935**	0.1200	*0.1424*	*0.1891*	0.0559	**0.1315**
	Flags	H. surv	Heart	Image Seg.	Landsat	Led
Gini	0.4794	0.2680	0.2037	0.0403	**0.1294**	0.3100
Info Gain	0.4639	**0.2647**	0.2444	0.0368	0.1361	0.3050
Twoing	*0.4845*	0.2745	*0.2481*	0.0485	0.1406	0.3200
C4.5	**0.4022**	*0.3070*	**0.2000**	0.0385	0.1324	*0.5869*
MEE	0.4691	**0.2647**	0.2444	*0.0589*	*0.1566*	**0.3000**
	Leukemia	LRS	Lymphog.	Mammog.	Monk	Mushrrom
Gini	0.1806	0.1450	*0.2754*	0.2084	0.1007	0.0004
Info Gain	*0.1944*	0.1450	*0.2754*	0.2157	0.1187	**0.0000**
Twoing	0.1667	**0.1431**	0.2061	0.2072	*0.1331*	**0.0000**
C4.5	**0.1389**	*0.2917*	0.2528	**0.2000**	0.1115	**0.0000**
MEE	0.1667	0.1638	0.2500	*0.2386*	**0.0989**	*0.0009*
	Ozone	Page blks	Parkinsons	Pen Digits	P. Diabetes	P. Gene
Gini	**0.0693**	0.0342	*0.1590*	0.0418	**0.2487**	0.2547
Info Gain	**0.0693**	0.0347	0.1487	**0.0357**	0.2695	*0.3208*
Twoing	0.0731	*0.0365*	0.1487	0.0378	0.2695	0.2642
C4.5	*0.0785*	**0.0281**	**0.1242**	0.0418	0.2578	0.2547
MEE	0.0704	0.0347	0.1436	*0.0666*	*0.3216*	**0.1698**
	Robot-1	Spect-Heart	Spectf-Heart	Swiss HD	Synth. Chart	Thyroid
Gini	0.2727	0.2022	0.2097	0.6117	0.1150	0.0844
Info Gain	0.2841	0.2060	**0.2060**	0.6083	0.0817	*0.1023*
Twoing	**0.1932**	*0.2210*	*0.2172*	*0.6250*	*0.1200*	0.0977
C4.5	*0.3500*	**0.1873**	0.2135	**0.5847**	0.0833	**0.0558**
MEE	0.2614	0.1985	**0.2060**	0.6083	**0.0617**	0.0977
	VA HD	Wdbc	Wpbc	Wine	Yeast	Zoo
Gini	*0.7527*	0.0650	**0.2371**	0.1067	0.4219	0.1683
Info Gain	*0.7527*	*0.0721*	0.2474	**0.0562**	0.4206	0.1683
Twoing	0.7419	0.0685	**0.2371**	0.0787	0.4381	0.1386
C4.5	0.7366	0.0650	*0.3144*	0.0899	**0.4077**	*0.3069*
MEE	**0.7097**	**0.0615**	**0.2371**	*0.1180*	*0.5335*	**0.1089**

	Gini	Info Gain	Twoing	C4.5	MEE	p
Wins	7	9	6	14	13	0.27
Losses	6	8	9	10	12	0.70

average tree size) and losses (highest average tree size) are also shown in Table 6.29 with the chi-square p. A significant difference is found relative to the equal distribution hypothesis, which is confirmed by other statistical tests.

Table 6.29 Tree size average (range) with wins (bold) and losses (italic).

	Arrythmya	Balance	Car	Clev. HD2	Clev.HD5	CNS
Gini	12.2 (10)	**26.6 (10)**	66.2 (38)	**6.8 (8)**	**5.6 (16)**	**1.6 (2)**
Info Gain	12.8 (8)	**26.6 (20)**	58.0 (34)	7.8 (8)	6.8 (8)	2.0 (2)
Twoing	**11.6 (14)**	27.6 (8)	66.0 (26)	8.8 (10)	5.8 (12)	2.4 (2)
C4.5	*37.8 (24)*	43.6 (12)	71.4 (12)	*19.2 (10)*	*34.8 (16)*	*6.2 (4)*
MEE	36.2 (10)	*90.6 (52)*	*115.0 (48)*	18.8 (12)	22.0 (42)	3.2 (4)
	Colon	Cork stop.	Credit	CTG	Dermatol.	E-coli
Gini	**3.0 (0)**	**5.0 (0)**	**3.0 (0)**	74.6 (46)	13.4 (6)	10.8 (12)
Info Gain	**3.0 (6)**	**5.0 (0)**	**3.0 (0)**	70.2 (52)	15.8 (6)	10.2 (14)
Twoing	**3.0 (4)**	**5.0 (0)**	**3.0 (0)**	*60.0 (30)*	*16.2 (2)*	11.0 (14)
C4.5	*5.8 (2)*	*5.8 (6)*	*24.8 (20)*	136 (26)	**13.0 (0)**	*17.2 (8)*
MEE	3.2 (2)	**5.0 (0)**	12.4 (24)	**56.8 (16)**	14.1 (2)	**9.4 (2)**
	Flags	H. surv	Heart	Image Seg.	Landsat	Led
Gini	12.8 (12)	*4.2 (28)*	11.2 (12)	*79.8 (58)*	**108.8 (80)**	22.0 (24)
Info Gain	**10.0 (14)**	**1.0 (0)**	10.8 (32)	55.2 (36)	131.2 (98)	**20.6 (8)**
Twoing	10.6 (20)	2.6 (8)	**8.4 (10)**	70.4 (74)	110.4 (98)	*25.6 (20)*
C4.5	*27.8 (6)*	*4.2 (6)*	17.2 (8)	59.8 (18)	*331.6 (58)*	22.0 (6)
MEE	20.0 (36)	**1.0 (0)**	*19.4 (14)*	**32.4 (8)**	125.6 (56)	23.0 (4)
	Leukemia	LRS	Lymphog.	Mammog.	Monk	Mushrrom
Gini	**3.0 (0)**	11.6 (4)	**6.0 (4)**	**5.8 (14)**	28.8 (28)	18.2 (8)
Info Gain	**3.0 (0)**	11.8 (6)	**6.0 (4)**	6.2 (6)	30.0 (24)	**16.2 (2)**
Twoing	**3.0 (0)**	**11.2 (2)**	6.4 (4)	6.2 (6)	**27.0 (24)**	18.8 (2)
C4.5	*3.8 (2)*	*28.8 (12)*	13.8 (8)	16.8 (4)	31.0 (14)	23.0 (0)
MEE	**3.0 (0)**	24.2 (6)	*17.6 (22)*	*24.6 (12)*	33.6 (28)	*42.6 (16)*
	Ozone	Page blks	Parkinsons	Pen Digits	P. Diabetes	P. Gene
Gini	**1.0 (0)**	23.4 (28)	5.4 (8)	336.8 (184)	6.0 (4)	7.8 (10)
Info Gain	**1.0 (0)**	23.4 (22)	7.6 (10)	327.6 (216)	6.6 (14)	**5.6 (8)**
Twoing	3.4 (24)	22.6 (24)	8.8 (18)	*371.2 (152)*	8.4 (36)	7.0 (8)
C4.5	*54.0 (34)*	*54.8 (16)*	*15.0 (4)*	271.4 (32)	*30.6 (28)*	*11.4 (6)*
MEE	**1.0 (2)**	**20.8 (6)**	**3.0 (0)**	**233.0 (60)**	**2.2 (4)**	8.6 (8)
	Robot-1	Spect-Heart	Spectf-Heart	Swiss HD	Synth. Chart	Thyroid
Gini	**8.2 (4)**	6.0 (14)	1.6 (6)	1.8 (8)	24.8 (20)	8.0 (6)
Info Gain	8.8 (4)	13.4 (24)	**1.0 (0)**	**1.0 (0)**	37.4 (18)	7.4 (8)
Twoing	9.4 (2)	**5.8 (20)**	2.2 (12)	1.6 (6)	*39.2 (24)*	*9.2 (12)*
C4.5	*10.4 (4)*	13.6 (4)	*28.4 (12)*	*17.8 (16)*	28.8 (6)	9.0 (4)
MEE	**8.2 (2)**	*23.2 (28)*	**1.0 (0)**	**1.0 (0)**	**22.0 (2)**	**5.0 (0)**
	VA HD	Wdbc	Wpbc	Wine	Yeast	Zoo
Gini	**5.8 (18)**	10.2 (14)	**1.0 (0)**	*11.8 (16)*	22.4 (22)	10.6 (6)
Info Gain	10.0 (18)	10.2 (14)	1.6 (6)	8.4 (2)	**22.0 (16)**	10.8 (4)
Twoing	8.6 (28)	11.0 (20)	**1.0 (0)**	7.8 (4)	17.8 (12)	**10.4 (2)**
C4.5	*26.0 (20)*	*15.4 (10)*	*13.8 (26)*	8.8 (2)	*135.2 (40)*	*11.0 (0)*
MEE	23.4 (12)	**5.2 (2)**	**1.0 (0)**	**6.6 (4)**	25.4 (10)	*11.0 (0)*

	Gini	Info Gain	Twoing	C4.5	MEE	p
Wins	15	15	11	1	20	0.00
Losses	3	0	5	27	9	0.00

Briefly, from the experimental evidence provided by the cited works [152, 153] one extracts the following conclusions:

- A comparable performance with respect to test error rates is observed for all algorithms; for instance, the Friedman statistical test (of the test error rates) didn't reject the hypothesis of equal distribution, with $p > 0.4$ in both works.

- The MEE algorithm produces, on average, smaller trees than the C4.5 algorithm [152, 153] and CART-IG [152] with better generalization and without significant sacrifice on performance. Denoting by $\hat{P}_{ed}$ and $\hat{P}_{et}$ respectively the mean training set error rate (resubstitution estimate) and the mean test set error rate (cross-validation estimate), the generalization was evaluated by computing $D = |\hat{P}_{ed} - \hat{P}_{et}|/\bar{s}$, with the pooled standard deviation $\bar{s}$. The Friedman test found a significant difference ($p \approx 0$) of the methods for the D scores with the post-hoc Dunn-Sidak statistical test revealing a significant difference between MEE vs C4.5 and vs CART-TWO as illustrated in Fig. 6.40.
- Judging from the tree size ranges [139], the MEE algorithm is significantly more stable than competing algorithms.
- The MEE algorithm is quite insensitive to pruning, at least when cost-complexity pruning is used. With this pruning method the solutions with or without pruning were found to be not significantly different, and as a matter of fact were coincident in many cases [152].

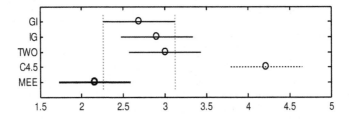

Fig. 6.40 Dunn-Sidak comparison intervals for the D scores.

Over-fitting of tree solutions to training sets is the reason why the pruning operation is always performed by tree design algorithms. It can be detected during the design phase by setting aside a test set and looking to its error rate during tree growing; over-fitting is then revealed by an inflected test error curve, going upward after a minimum. Figure 6.41 shows, for the consecutive tree levels, the mean training set and test set error rates ($\pm$standard deviation) in 20 experiments of a MEE tree designed for the Ionosphere dataset [13]. The training set used 85% of randomly chosen cases and testing was performed in the remaining 15%. There is no evidence of over-fitting in this case. According to [152] in only 1/8 of the datasets the MEE design revealed mild over-fitting symptoms (in the last one or two levels). In that work a comparison between test set error rates of pruned and unpruned solutions is also reported in detail; no statistical significant difference ($p = 0.41$) was found between the two groups of designed solutions.

Finally, a comparison of computation times is also presented in [152]. When the only difference among the algorithms is the implementation of the splitting criteria, then the MEE tree algorithm may take substantially less time to

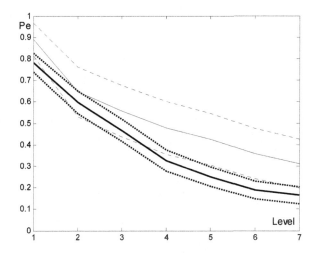

Fig. 6.41 Mean (solid line) and mean ± std (dashed line) of the training set error (black, thick line) and test set error (grey, thin line) in 20 experiments of trees designs for the Ionosphere dataset.

design a tree for a not too large number of classes. As a matter of fact the time needed to compute a MEE split can be ten to hundred times smaller than to compute other types of splits. When the number of classes is large (say, above 8), even with the provisions discussed in Sect. 6.6.1.3, a substantial increase in computation time may occur.

Appendix A
Maximum Likelihood and Kullback-Leibler Divergence

A.1 Maximum Likelihood

Let us consider a set $D_n = \{x_i; i = 1, \ldots, n\}$ of n observations of a random variable X distributed according to a PDF belonging to a family $\{p(x; \theta)\}$ with unknown parameter vector $\theta \in \Theta$. The maximum likelihood (ML) method provides an estimate of θ that best supports the observed set D_n in the sense of maximizing the likelihood $p(D_n|\theta)$, which for any θ is given by the joint density value

$$p(D_n|\theta) = p(x_1, x_2, \ldots, x_n|\theta) . \tag{A.1}$$

Let us assume that the observations are i.i.d. realizations of $p(x; \theta)$. The likelihood can then be written as

$$p(D_n|\theta) = p(x_1, x_2, \ldots, x_n|\theta) = \prod_{i=1}^{n} p(x_i; \theta) . \tag{A.2}$$

Note that $p(D_n|\theta)$ is a function of the parameter vector θ and, as a function of θ, it is *not* a probability density function (for instance its integral may differ from 1). Since the logarithm function is monotonic and given the exponential form of many common distributions, it is usually more convenient to maximize the log-likelihood instead of (A.2). We then search for

$$\hat{\theta} = \arg\max_{\theta \in \Theta} \mathcal{L}(\theta|D_n), \text{ with } \mathcal{L}(\theta|D_n) = \sum_{i=1}^{n} \ln p(x_i; \theta) . \tag{A.3}$$

For discrete distributions one can still apply formula (A.3) interpreting the $p(x_i; \theta)$ as PMF values. The method is quite appellative and has excellent mathematical properties especially for large n (see e.g., [136]).

A.2 Kullback-Leibler Divergence

The Kullback-Leibler (KL) divergence (also called relative entropy) is a discrepancy measure between two probability distributions p and q. For discrete distributions, with p and q representing PMFs, the KL divergence is denoted and defined as:

$$D_{KL}(p||q) \equiv D_{KL}(p(x)||q(x)) = \sum_x p(x) \ln \frac{p(x)}{q(x)} . \qquad (A.4)$$

For continuous distributions p and q represent densities and the summation is substituted by an integral.

We may also write $D_{KL}(p||q)$ as

$$\begin{aligned} D_{KL}(p||q) &= - \sum_x p(x) \ln q(x) + \sum_x p(x) \ln p(x) \\ &= H_S(p,q) - H_S(p) , \end{aligned} \qquad (A.5)$$

where $H_S(p)$ is the (Shannon) entropy of the distribution p and $H_S(p,q)$ is the cross-entropy between p and q.

Note that the KL divergence is not a metric distance because it does not satisfy the symmetry property, $D_{KL}(p||q) \neq D_{KL}(q||p)$, nor the triangle inequality. However, it has some interesting properties, namely $D_{KL}(p||q) \geq 0, \forall p(x), q(x)$, and $D_{KL}(p||q) = 0$ iff $p(x) = q(x), \forall x$.

From (A.4) we observe that

$$D_{KL}(p||q) = \mathbb{E}_p \left[\ln \frac{p(x)}{q(x)} \right] , \qquad (A.6)$$

with $\mathbb{E}_p$ denoting an expectation relative to the distribution p. We are then able to compute the empirical estimate (resubstitution estimate)

$$\hat{D}_{KL}(p||q) = \frac{1}{n} \sum_{i=1}^{n} \ln \frac{p(x_i)}{q(x_i)} . \qquad (A.7)$$

Since $\hat{D}_{KL}(p||q)$ is an empirical measure of the discrepancy between $p(x)$ and $q(x)$ its minimization can be applied to finding a distribution q approximating another distribution p.

A.3 Equivalence of ML and KL Empirical Estimates

Let us assume we use (A.7), instead of the ML method, for the estimation of the parameter vector θ mentioned in Sect. A.1. The distribution of the i.i.d. x_i is $p(x; \theta_0)$ with θ_0 unknown. We estimate it by attempting to find

$$\hat{\theta} = \arg\min_{\theta \in \Theta} \hat{D}_{KL}(p(x;\theta_0)||p(x;\theta))$$

with

$$\hat{D}_{KL}(p(x;\theta_0)||p(x;\theta)) = \frac{1}{n}\sum_{i=1}^{n}\ln\frac{p(x_i;\theta_0)}{p(x_i;\theta)} \ . \tag{A.8}$$

But since (A.8) can be written as

$$\frac{1}{n}\sum_{i=1}^{n}\ln p(x_i;\theta_0) - \frac{1}{n}\sum_{i=1}^{n}\ln p(x_i;\theta) \ , \tag{A.9}$$

and the left term of (A.9) doesn't depend on θ, the estimated parameter vector $\hat{\theta}$ obtainable by minimizing the empirical KL estimate (A.8) is the same as the one expressed by (A.3), obtainable by the ML method.

Appendix B
Properties of Differential Entropy

B.1 Shannon's Entropy

$$H_S(X) = -\int_X f(x) \ln f(x) dx = -\mathbb{E}[lnf(X)] . \tag{B.1}$$

A list of important properties of Shannon's entropy [48, 184, 62] is:

1. $H_S(X) \in \,]-\infty, \ln \|X\|]$, where $\|X\|$ is the support length of X. The equality $H_S(X) = \|X\|$ holds for a uniform distribution in a bounded support. The minimum value $(-\infty)$ corresponds to a sequence of continuous Dirac-δ functions (Dirac-δ comb).
2. Invariance to translations: $H_S(X + c) = H_S(X)$ for a constant c.
3. Change of scale: $H_S(aX) = H_S(X) + \ln |a|$ for a constant a. If X is a random vector, $H_S(\mathbf{A}X) = H_S(X) + \ln |\det \mathbf{A}|$.
4. Conditional entropy: $H_S(X|Y) = -\mathbb{E}[\ln f(X|Y)] \leq H_S(X)$ with equality iff X and Y are independent.
5. Sub-additivity for joint distributions: $H_S(X_1, \ldots, X_n) = \sum_{i=1}^{n} H_S(X_i|X_1, \ldots, X_{i-1}) \leq \sum_{i=1}^{n} H_S(X_i)$, with equality (additivity) only if the r.v.s are independent.
6. Bijective transformation $Y = \varphi(X)$: $H_S(Y) = H_S(X) - E_X[\ln |J_\varphi(Y)|]$, where $J_\varphi(Y) = \left[\frac{\partial \varphi^{-1}(y_i)}{\partial y_k}\right]$, $i, k = 1, \ldots, d$, is the Jacobian of the transformation. Note that this implies properties 2 and 3, and also the invariance under an orthonormal transformation $Y = \mathbf{A}X$, with $|\mathbf{A}| = 1$.
7. If a random vector X with support in $\mathbb{R}^n$ has covariance matrix $\mathbf{\Sigma}$, then $H_S(X) \leq \frac{1}{2}\ln[(2\pi e)^n |\mathbf{\Sigma}|]$. The equality holds for the multivariate Gaussian distribution.
8. Let $X_1, \ldots, X_n$ be independent r.v.s with densities and finite variances. Then,

$$e^{2H_S(X_1+\ldots+X_n)} \geq \sum_{i=1}^{n} e^{2H_S(X_i)} . \tag{B.2}$$

This last result is of a monotonicity theorem first proved by A. Stam in 1959 [222]. It is also presented and discussed in [50]. Sometimes this (and other) results are expressed in terms of the *entropy power* defined as $N(X) = \frac{1}{2\pi e}e^{2H_S(X)}$. We now present a Corollary of this theorem justifying larger Shannon entropy of Gaussian smoothed distributions (see Chap. 3).

Corollary B.1. *Let* X *and* Y *be independent continuous random variables and* $f_Z = f_X \otimes f_Y$. *If* Y *is Gaussian distributed with variance* h^2, *then*

$$H_S(Z) = H_S(X + Y) \geq H_S(X) . \tag{B.3}$$

Proof. From the monotonicity theorem we have:

$$N(X + Y) \geq N(X) + N(Y) . \tag{B.4}$$

Since Y is a Gaussian r.v. we also have $N(Y) = \frac{1}{2\pi e}e^{2H(Y)} = \frac{1}{2\pi e}e^{2\ln(h\sqrt{2\pi e})} = h^2 > 0$, and the above result follows. $\qquad\square$

B.2   Rényi's Entropy

$$H_{R_\alpha}(X) = \frac{1}{1-\alpha}\ln\int_X f^\alpha(x)dx = \frac{1}{1-\alpha}\ln\mathbb{E}[f^{\alpha-1}(x)],\ \alpha \geq 0,\ \alpha \neq 1 \tag{B.5}$$

A list of important properties of Rényi's entropy [47, 62] is:

1. H_{R_α} can be positive or negative, but H_{R_2} is non-negative with minimum value (0) corresponding to a Dirac-δ comb.
2. Invariance to translations: $H_{R_\alpha}(X + c) = H_{R_\alpha}(X)$ for a constant c.
3. Change of scale: $H_{R_\alpha}(aX) = H_{R_\alpha}(X) - \ln|a|/(1-\alpha)$ for a constant a.
4. With the conditional entropy defined similarly as for the Shannon entropy (Property 4 of Sect. B.1), the inequality $H_{R_\alpha}(X|Y) \leq H_{R_\alpha}(X)$ only holds for $\alpha \leq 1$ and $f(x) \leq 1$ in the whole support.
5. $H_{R_\alpha}(X_1,\dots,X_n) = \sum_{i=1}^n H_{R_\alpha}(X_i)$ for independent random variables.
6. For $\alpha = 2$ and univariate distributions with finite variance σ^2, the maximizer of Rényi's quadratic entropy is $\frac{1}{\sqrt{5\pi}\sigma}\frac{\Gamma(5/2)}{\Gamma(2)}\left(1 - \frac{x^2}{5\sigma^2}\right)$ with support $|x| \leq \sqrt{5}\sigma$. The general formulas of the maximizing density of the α-Rényi entropy are given in [47]. Note that the Rényi entropy of the univariate normal distribution [162] is $H_{R_\alpha}(g(x; \mu, \sigma)) = \ln(\sqrt{2\pi}\sigma) - \frac{1}{2}\ln\alpha/(1-\alpha)$; therefore, $H_{R_2}(g(x; \mu, \sigma)) = \ln(2\sigma\sqrt{\pi})$.
7. $H_{R_\alpha}(X) \xrightarrow{\alpha\to1} H_S(X)$.

Note that a result similar to the one in Corolary B.1 for Rényi's quadratic entropy is a trivial consequence of properties 1 and 5.

Appendix C
Entropy and Variance of Partitioned PDFs

Consider a PDF $f(x)$ defined by a weighted sum of functions with disjoint supports,

$$f(x) = \sum_i a_i f_i(x) \;, \tag{C.1}$$

such that

1. Each $f_i(x)$ is a PDF with support D_i;
2. $D_i \cap D_j = \varnothing, \forall i \neq j$;
3. The support of f is $D = \cup_i D_i$;
4. $\sum_i a_i = 1$.

We call such an $f(x)$ a *partitioned* PDF. From (C.1), and taking the above conditions into account, the Shannon (differential) entropy of $f(x)$ is expressed as

$$
\begin{aligned}
H_S(f) &= -\int_D \left[\sum_k a_k f_k(x) \right] \ln \left[\sum_k a_k f_k(x) \right] dx = \\
&= -\sum_i \int_{D_i} \left[\sum_k a_k f_k(x) \right] \ln \left[\sum_k a_k f_k(x) \right] dx = \\
&= -\sum_i \int_{D_i} a_i f_i(x) \left[\ln a_i + \ln f_i(x) \right] dx = \\
&= -\sum_i a_i \int_{D_i} f_i(x) \ln f_i(x) dx - \sum_i a_i \ln a_i \;.
\end{aligned} \tag{C.2}
$$

We then have

$$H_S(f) = \sum_i a_i H_S(f_i) - \sum_i a_i \ln a_i \;. \tag{C.3}$$

Thus, the Shannon entropy of f is a weighted sum of the entropies of each component f_i plus the entropy of the PMF corresponding to the weighting factors. This is the continuous counterpart of the partition invariance property of discrete Shannon entropy (see e.g. [62]).

Rényi's quadratic (differential) entropy of a partitioned PDF is expressed as

$$H_{R_2}(f) = -\ln\left[\sum_i a_i^2 \int_{D_i} f_i^2(x)dx\right] \; ; \tag{C.4}$$

that is, $H_{R_2}(f)$ is not decomposable as Shannon's counterpart. Nevertheless, as the minimization of $H_{R_2}(f)$ is equivalent to the maximization of $V_{R_2} = \exp(-H_{R_2})$, we may use the decomposition of the information potential V_{R_2} which is readily expressed as

$$V_{R_2}(f) = \sum_i a_i^2 V_{R_2}(f_i) \; . \tag{C.5}$$

The variance of f can also be decomposed as

$$V[f] = \sum_i a_i V[f_i] + \sum_i a_i(\mu_i - \mu)^2 \; , \tag{C.6}$$

where $\mu(\mu_i)$ is the expected value of $f(f_i)$.

Appendix D
Entropy Dependence on the Variance

When analyzing the MSE and EE risk functionals in a comparative way, one often wishes to know how the Shannon entropy changes with the variance (denoted in this Appendix simply as H and V, respectively). We shall see in Sect. D.2 that many unimodal parametric PDFs, $f(x; \alpha)$, have an entropy $H(\alpha)$ which changes with the variance $V(\alpha)$ in an increasing way but with decreasing derivative [219]. In order to show this we first need to introduce the notion of saturating functions.

D.1 Saturating Functions

Definition D.1. A real continuous function $f(x)$ with domain $\mathbb{R}^+$ is an *up-saturating* (*down-saturating*) function — denoted USF (DSF) — if it is strictly concave (convex) and increasing (decreasing). □

Remarks:

1. Let us recall that $f(x)$ is a strictly concave function if $\forall t \in]0, 1[$ and $x_1 \neq x_2$ then $f(tx_1 + (1-t)x_2) > tf(x_1) + (1-t)f(x_2)$. Function f is strictly convex if $-f$ is strictly concave.
2. One could define saturating functions for the whole real line. They are here restricted to $\mathbb{R}^+$ for simplicity and because variance is a non-negative quantity.
3. $\sqrt{x}$ and $\ln(x)$ are obvious USFs. The functions digamma, $\psi(x) = \frac{d\Gamma(x)}{dx}$, and trigamma, $\psi_1(x) = \frac{d\psi(x)}{dx}$, to be used below, are USF and DSF, respectively.

Properties. The following are obvious properties of saturating functions:

i If f is DSF $-f$ is USF and vice-versa.
ii If f is a saturating function and $a \in \mathbb{R}$ is a constant, $f + a$ is a saturating function of the same type (i.e., both either USF or DSF).

iii If f is a saturating function and $a \neq 0$ is a constant, af is a saturating function; of the same type if $a > 0$ and of different type otherwise.

iv If $f(x)$ is a saturating function and $a > 0$ is a constant, $f(ax)$ is a saturating function of the same type.

v If f_1 and f_2 are USFs of the same variable and with domains D_1 and D_2, $f_1 + f_2$ defined on $D_1 \cap D_2 \neq \emptyset$ is USF. A similar result holds for DSFs (yielding a DSF).

Remarks:

1. The above properties (i) through (iv) are useful when one has to deal with annoying normalization factors. Let us suppose a PDF $f(x) = kg(x)$, where k is the normalization factor. We have $H = -\int f \ln f = -kG - k \ln k$ with $G = \int g \ln g$. Let us further suppose that $V = av$, where $a > 0$ is a constant. We then have the following implications with annotated property: $G(v)$ DSF $\overset{iv}{\Rightarrow}$ $G(V)$ DSF $\overset{i}{\Rightarrow}$ $-G(V)$ USF $\overset{iii}{\Rightarrow}$ $-kG(V)$ USF $\overset{ii}{\Rightarrow}$ $H(V)$ USF.

2. By property (ii) if $H(V)$ is USF then H is also USF in terms of the second-order moment, i.e., the "MSE risk".

Proposition D.1. *If $f(x)$ is a saturating function and $g(x)$ is USF, with the codomain of g contained in the support of f, then $h(x) = f(g(x))$ is a saturating function of the same type of $f(x)$.*

Proof. Let us analyze the up-saturating case of $f(x)$ and set $y = g(x)$. Since g is strictly concave we have for $t \in]0,1[$ and $x_1 \neq x_2$, $g(tx_1 + (1-t)x_2) > ty_1 + (1-t)y_2$. Therefore, because f is a strictly increasing function, $h(tx_1 + (1-t)x_2) > f(ty_1 + (1-t)y_2)$, the strict concavity of f implying $h(tx_1 + (1-t)x_2) > tf(y_1) + (1-t)f(y_2)$. Furthermore, since the composition of two strictly increasing functions is a strictly increasing function, h is USF. The result for a down-saturating $f(x)$ is proved similarly. $\square$

Remark: The preceding Proposition allows us to say that if entropy is a saturating function of the standard deviation, σ, it will also be a saturating function of the variance, V, since $\sigma = \sqrt{V}$ is USF. The converse is not true. For instance, $f(V) = V^{0.7}$ is USF of V but not of $\sigma (f(\sigma) = \sigma^{1.4})$.

Proposition D.2. *A differentiable USF (DSF) $f(x)$, has a strictly decreasing positive (increasing negative) derivative and vice-versa.*

Proof. We analyze the up-saturating case. Since $f(x)$ is strictly concave, we have for $t \in]0,1[$, $(1-t)f(x) + tf(x+t) < f((1-t)x + t(x+t)) = f(x+t^2)$; therefore, $\frac{f(x+t) - f(x)}{t} < \frac{f(x+t^2) - f(x)}{t^2}$, and since $f(x)$ is strictly increasing the strict decreasing positive derivative follows. The down-saturating case is proved similarly. $\square$

Remark: The preceding Proposition is sometimes useful to decide whether the entropy, H, is a saturating function of the variance, V, by looking into dH/dV.

D.2 PDF Families with Up-Saturating $H(V)$

Many unimodal parametric PDF families exhibit an entropy (always understood here as Shannon's entropy) which is an up-saturating function of the variance. This means that the entropy is also increasing with the variance, but beyond the $dH/dV = 1$ point the entropy grows slower than the variance, exhibiting a saturation phenomenon.

The parametric PDFs analyzed in this section are those included in PDF lists published in the literature (see, namely, [121,122,28]) and the web, as well as those belonging to the Pearson system of distributions. A non-exhaustive list of PDFs with USF $H(V)$ is presented in Sect. D.2.2. The PDFs of the Pearson system are analyzed in Sect. D.2.1.

For many PDFs with known variance and entropy formulas, it is possible to elicit whether or not the entropy is an up-saturating function of the variance by applying the results of Sect. D.1. In more difficult cases one may have to numerically compute $H(V)$ and inspect the respective graphs. When the PDF family has only one parameter, say $f(x; \alpha)$, our analysis is on $H(V(\alpha))$; for more than one parameter, the results we present correspond to varying one of the parameters, setting the others to constant values.

Of particular interest to us is the PDF family defined by formula (3.43), corresponding to the perceptron error with Gaussian inputs and activation function *tanh*. (Note that for perceptrons with outputs obtained by applying an activation function to a weighted sum of the inputs, the Gaussianity of the inputs is not a stringent condition if the inputs are independently distributed.) No entropy formula is available for this PDF, which we shall call the density of the *tanh-neuron distribution*, rewritten for $t = 1$ (therefore, for $y = 1 - e$) as $f(y; a, b) = K \exp(-(\operatorname{atanh}(y) - a)^2/b) / ((1 - y)(1 + y))$, with a governing the mean and b the variance. K is a normalization factor. Figure D.1a shows this PDF for fixed a and three values of b (therefore, of the variance). By numerical computation one can confirm that the entropy is indeed an up-saturating function of V as shown in Fig. D.1b.

D.2.1   The Pearson System

The Pearson system of distributions [121, 176, 239] has densities defined by

$$\frac{df(x)}{dx} = \frac{m - x}{a + bx + cx^2} f(x) \,, \tag{D.1}$$

with solutions

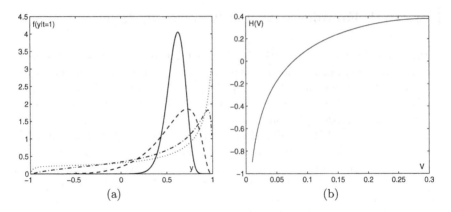

Fig. D.1 The tanh-neuron PDF for $a = 0.7$: a) PDFs for $b = 0.05$ (solid), $b = 0.3$ (dashed), $b = 1.2$ (dash-dot), $b = 2$ (dotted); b) $H(V)$ for $b \in [0.05, 2]$.

$$f(x) = K(a + bx + cx^2)^{-1/2c} \exp\left(\frac{(b + 2cm)\operatorname{atan}((b + 2cx)/\sqrt{4ac - b^2})}{c\sqrt{4ac - b^2}}\right).$$

$$(\text{D.2})$$

From (D.1) and (D.2) one obtains several families of distributions by imposing restrictions on the parameters a, b, c, and m, which control the shape of f. The roots of $a + bx + cx^2 = 0$ define distribution types. For some types one has to restrict the support of f in order to guarantee $f > 0$. K is a normalization constant guaranteeing $\int f = 1$.

The Pearson system is extraordinarily rich, in the sense that it includes many well-known distributions, as we shall see, and also allows accurate modeling of complex univariate PDFs of real-world datasets (see e.g., [10]).

Besides the Gauss distribution which is a special case of (D.2), there are twelve Pearson types of density families, depending on $\Delta = b^2 - 4ac$ and conditions on a, b, and c. For several types there are no known formulas for the variance and entropy (no known explicit integral), except sometimes for particular subtypes. Many entropy formulas can be found in [133] and [162] (in this last work the Rényi entropies are also presented). Others can be computed with symbolic mathematical software. The variance and entropy of the Pearson types are as follows:

Type 0 : $\Delta = 0$ and $b = c = 0$, $a > 0$.
Corresponds to the Gauss distribution family with mean $-m$ and $\sigma = \sqrt{a}$. The entropy $H(\sigma) = \ln(\sigma\sqrt{2\pi e})$ is a USF of V.

Type I: $\Delta > 0$, real roots a_1 and a_2 of opposite signs ($a_1 < 0 < a_2$), and $x \in [a_1, a_2]$.
There are no known formulas for the variance and entropy of the general solution, $f(x) = K(1 + x/a_1)^{m_1}(1 - x/a_2)^{m_2}$, $-a_1 < x < a_2$, $m_1, m_2 > -1$. However, a particular subtype is the generalized Beta distribution

family (of the first kind), defined for $x \in [0, 1]$ and $\alpha, \beta > 0$ as $f(x) = x^{\alpha-1}(1-x)^{\beta-1}/B(\alpha, \beta)$. Variance and entropy for this subtype are given in Sect. D.2.2. Beta densities can be symmetric, asymmetric, convex or concave. Under certain conditions on (α, β) the entropy is *not* an up-saturating function of the variance, or even not a saturating function at all. For instance, for $\alpha = \beta$ and $\alpha < 1$ (U-shaped densities) the variance increases with decreasing entropy! For $\alpha \gg \beta$ one can write $V \cong \beta/\alpha^2$ and $H(\alpha, \beta) \cong \ln \Gamma(\beta) + \psi(\beta)$, which is then an up-saturating function of V. The same holds for $\beta \gg \alpha$ (both V and H are symmetric in α, β). Numerical computation shows that it is enough that one of the parameters is less than one-half of the other for the result to hold.

Type II: $\Delta > 0$, $b = 0$.
A version of type I corresponding to the density $f(x) = K(1 - x^2/a^2)^m$, $-a < x < a$, $a > 0$, $m > -1$. Variance and entropy are given in Sect. D.2.2. For fixed m a family of PDFs with the same basic shape is obtained and the entropy is in this case an up-saturating function of the variance. For fixed a the shape of the PDFs varies a lot with m and the entropy is not a saturating function of the variance. Note that the uniform density is a special case of this type and its entropy is a USF of V.

Type III: $\Delta > 0$ with $c = 0$, a, $b \neq 0$.
There are no known formulas for the variance and entropy of the general solution, $f(x) = K(1+x/a)^{ma} \exp(-mx)$, $-a < x < \infty$, a, $m > 0$. A subtype is the Gamma family, $f(x) = x^{k-1}e^{-x/\theta}/(\theta^k \Gamma(k))$ for $x > 0$, k (shape), θ(scale) > 0, with $V(k, \theta) = k\theta^2$ and $H(k, \theta) = k + \ln \theta + \ln \Gamma(k) + (1-k)\psi(k)$. The gamma distribution family is often used for PDF modeling. Particular cases of the Gamma family are the exponential, the Erlang, the chi-square and the Maxwell-Boltzmann families. For fixed k the entropy is $\ln(\sqrt{V})$ plus a constant, which is an obvious up-saturating function. For fixed θ the entropy can be written in terms of $v = v/\theta^2$ as $H(v, \theta) = v + \ln \theta + \ln \Gamma(v) + (1-v)\psi(v)$. Since $dH/dv = 1 + (1 - v)\psi_1(\nu)$ is a decreasing function, the entropy is also in this case a USF of V.

Type IV: $\Delta < 0$.
The general solution is $f(x) = K(1+(x/a)^2)^{-m} \exp(-k\mathrm{atan}(x/a))$, a, $k > 0$, which describes a family of long-tailed distributions, where a is a scale parameter, k an asymmetry parameter and m determines the tails (long tails for small m). The Pearson Type IV is used in several areas (namely economics and physics) whenever one needs to model empirical distributions with long tails.

The best way to mathematically handle the Pearson Type IV family is by performing the variable transformation $\tan \theta = x/a$ [243]. The value of K is defined only for $m \geq 1/2$ and is given by $K = 1/(aF(r, k))$ with $r = 2m - 2$ and $F(r, k) = \int_{\pi/2}^{\pi/2} e^{-k\theta} \cos^r(\theta)d\theta$ (closed form expression only for integer r). The variance is given by $V = a^2(r^2 + k^2)/[r^2(r - 1)]$ for $m > 3/2$.

There is no known closed form of the entropy, which can be written as $H = -I(r, k)/F(r, k) + \ln a + \ln F(r, k)$ ($I(r, k)$ is defined in D.2.2). Therefore, in terms of a, $H(V)$ is clearly USF. Setting $k = 1$, numerical computation leads to the conclusion that $H(V)$ is also USF in terms of r (and therefore of m, the parameter controlling the tails).

Student's t-distribution is the special symmetrical case ($k = 0$) of this family. It is easy to check that the entropy of the Student's t distribution is a USF of V.

Type V: $\Delta = 0$.
Corresponds to the Inverse Gamma family $f(x) = \beta^\alpha \frac{1}{x^{\alpha+1}} \exp(-\beta/x)/\Gamma(\alpha)$, with $x > 0$, α (shape), β (scale) > 0. The variance and the entropy are given in Sect. D.2.2. Note that in order for $f(x)$ to have variance α must be larger than 2. For fixed α, the entropy is $\ln(\sqrt{V})$ plus a constant, which is an obvious USF. For fixed β, numerical computation shows that the entropy is also a USF of V.

Type VI: $\Delta > 0$ and $x \geq a_2$, the larger root.
Corresponds to $f(x) = Kx^{-q_1}(x-a)^{q_2}$, with $q_1 < 1$, $q_2 > -1$, $q_1 > q_2 - 1$, and $x \in] a, \infty]$. There is no known entropy formula for the general solution. A particular subtype is the Beta distribution of the second kind (also known as Beta prime distribution), with $f(x) = x^{\alpha-1}(1+x)^{-\alpha-\beta}/B(\alpha, \beta)$ for $x \geq 0$. The variance is given in Sect. D.2.2; there is no known formula of the entropy. Numerical computation shows that $H(V)$ is a USF function both in terms of α and β.

Another special subtype is the F distribution family, whose variance and entropy are given in Sect. D.2.2. Numerical computation shows that the entropy is a USF of V, in relation to either of the two parameters (degrees of freedom) of this family.

Type VII: $\Delta > 0$, $b = 0$, $c > 0$.
Corresponds to $f(x) = K(1+x^2/a^2)^{-m}$, with $a \neq 0$, $m > 1/2$. A particular case is the Student's t-density already analyzed in type IV. The entropy of Type VII distributions is a USF of the variance for fixed a or m.

Type VIII: $\Delta > 0$ and $a_1 = a_2 = -m$.
Corresponds to $f(x) = K(1 + x/a)^{-m}$, with $m > 1$ and $x \in] - a, 0]$. These are asymmetric distributions, with hyperbolic shape sharply peaked for high m. Setting w.l.o.g. $a = 1$ (a simply controls where the peak occurs) and $x \in [-1 + \epsilon, 0]$, with $\epsilon \in]0, 1[$ (the area is infinite for $\epsilon = 0$), one obtains a family of distributions with variance defined for $m > 2$. Numerical computation shows that the entropy of the family is then an up-saturating function of the variance.

Type IX: $\Delta > 0$ and $a_1 = a_2 = m$.
A version of Type VIII, with $f(x) = K(1 + x/a)^m$, $m > -1$ and $x \in] - a, 0]$. The area is finite for $m \geq 0$, which affords simpler formulas for

the variance and entropy (given in Sect. D.2.2). The entropy is a USF of the variance.

Type X: $\Delta = 0$ and $b = c = 0$.
Corresponds to the generalized exponential family, $f(x) = \sigma e^{-(x-m)/\sigma}$, with $x \in [m, \infty[$ and $\sigma > 0$. As already noted for type III, the entropy is an up-saturating function of the respective variance (H is insensitive to m; see also remark on property (ii) of Sect. D.1).

Type XI: $\Delta = 0$ and $b = c = m = 0$.
Corresponds to $f(x) = Kx^{-\alpha}$, with $\alpha > 1$, $x \in [\epsilon, \infty[, \epsilon > 0$. Variance (defined for $\alpha > 3$) and entropy formulas are easily obtained (see Sect. D.2.2). The entropy is a USF of the variance. A special case of this type is the Pareto family of distributions.

Type XII: is essentially a version of type I with $m_1 = m_2 = m$, $|m| < 1$, and contains as special case the Beta distributions with $\alpha = \beta$. What we said for type I also apply here.

D.2.2 List of PDFs with USF $H(V)$

The following list covers by alphabetical order most of the PDF families described in the literature, namely those having closed form integrals for the variance and the entropy. In a few cases where no formulas were available we had to resort to numerical computation. The list is by no means intended to be an exhaustive list of PDFs with up-saturating $H(V)$ (where up-saturating is understood in relation to all parameters of interest except when stated otherwise).

1. **Beta** distribution (also known as Beta distribution of the first kind).
   PDF: $f(x; \alpha, \beta) = \frac{x^{\alpha-1}(1-x)^{\beta-1}}{B(\alpha,\beta)}$; $x \in [0, 1]$; α, $\beta > 0$.
   V : $\frac{\alpha\beta}{(\alpha+\beta)^2(\alpha+\beta+1)}$.
   H : $\ln(B(\alpha, \beta)) - (\alpha - 1)\psi(\alpha) - (\beta - 1)\psi(\beta) + (\alpha + \beta - 2)\psi(\alpha + \beta)$.
   Comment: A special case of Pearson Type I, corresponding to symmetric distributions in α, β. $H(V)$ is USF only for $\alpha \gtrsim 2\beta$ or $\beta \gtrsim 2\alpha$.

2. **Beta Prime** distribution (also known as Beta distribution of the second kind).
   PDF: $f(x; \alpha, \beta) = \frac{x^{\alpha-1}(1+x)^{-\alpha-\beta}}{B(\alpha,\beta)}$; $x > 0$; α, $\beta > 0$.
   V : $\frac{\alpha(\alpha+\beta-1)}{(\beta-2)(\beta-1)^2}$; $\beta > 2$.
   H: (no known formula).
   Comment: a special case of Pearson Type VI.

3. **Chi** distribution.
   PDF: $f(x; k) = \frac{2^{1-k/2}x^{k-1}e^{-x^2/2}}{\Gamma(k/2)}$; $x \geq 0$, $k > 0$.

$V:\ k-2\left(\frac{\Gamma((k+1)/2)}{\Gamma(k/2)}\right)^2.$

$H:\ \ln\Gamma\left(\frac{k}{2}\right)+\frac{1}{2}\left(k-\ln 2-(k-1)\psi\left(\frac{k}{2}\right)\right).$

4. **Chi-Square** distribution.

PDF: $f(x;k)=\frac{1}{2^{k/2}\Gamma(k/2)}x^{k/2-1}e^{-x/2};\ x\geq 0,\ k>0.$

$V:\ 2k.$

$H:\ \frac{k}{2}+\ln 2+\ln\left(\frac{k}{2}\right)+\left(1-\frac{k}{2}\right)\psi\left(\frac{k}{2}\right).$

Comment: a special case of the Gamma distribution.

5. **Erlang** distribution.

PDF: $f(x;k,\lambda)=\frac{\lambda^k x^{k-1}e^{-\lambda x}}{(k-1)!};\ x\geq 0;\ k,\ \lambda>0.$

$V:\ k/\lambda^2.$

$H:\ (1-k)\psi(k)+\ln\frac{\Gamma(k)}{\lambda}+k.$

Comment: a special case of the Gamma distribution.

6. **Exponential** distribution.

PDF: $f(x;\lambda)=\lambda e^{-\lambda x};\ x\geq 0;\ \lambda>0.$

$V:\ 1/\lambda^2.$

$H:\ 1-\ln\lambda.$

Comment: a special case of the Gamma distribution; $H(V)$ is also USF for the generalized exponential distribution (see Pearson type X).

7. **F** distribution.

PDF: $f(x;d_1,\ d_2)=\frac{1/x}{B\left(\frac{d_1}{2},\frac{d_2}{2}\right)}\sqrt{\frac{(d_1 x)^{d_1}d_2^{d_2}}{(d_1 x+d_2)^{d_1+d_2}}};\ x\geq 0,\ d_1,\ d_2>0.$

$V:\ \frac{2d_2^2(d_1+d_2-2)}{d_1(d_2-2)^2(d_2-4)};\ d_2>4.$

$H:\ \ln\left(\frac{d_1}{d_2}B\left(\frac{d_1}{2},\frac{d_2}{2}\right)\right)+\left(1-\frac{d_1}{2}\right)\psi\left(\frac{d_1}{2}\right)-\left(1+\frac{d_2}{2}\right)\psi\left(\frac{d_2}{2}\right)+$
$+\frac{d_1+d_2}{2}\psi\left(\frac{d_1+d_2}{2}\right).$

Comment: a special case of Pearson Type VI.

8. **Gamma** distribution.

PDF: $f(x;k,\theta)=x^{k-1}e^{-x/\theta}/(\theta^k\Gamma(k));\ x>0;\ k,\ \theta>0.$

$V:\ k\theta^2.$

$H:\ k+\ln\theta+\ln\Gamma(k)+(1-k)\psi(k).$

9. **Gauss** distribution.

PDF: $f(x;\mu,\sigma)=\frac{1}{\sqrt{2\pi}\sigma}e^{-\frac{(x-\mu)^2}{2\sigma^2}};\ x\in\mathbb{R}.$

$V:\ \sigma^2.$

$H:\ \ln(\sigma\sqrt{2\pi e}).$

10. **Generalized Gaussian** distribution.

PDF: $f(x;\alpha,\beta,\mu)=\frac{\beta}{2\alpha\Gamma(1/\beta)}\exp\left(-(|x-\mu|/\alpha)^\beta\right);\ x\in\mathbb{R};\ \alpha,\ \beta>0.$

$V:\ \frac{\alpha^2\Gamma(3/\beta)}{\Gamma(1/\beta)}\ .$

$H:\ \frac{1}{\beta}-\ln\left(\frac{\beta}{2\alpha\Gamma(1/\beta)}\right).$

11. **Gumbel** distribution.

    PDF: $f(x; \beta) = \frac{1}{\beta} e^{-x/\beta} e^{-e^{-x/\beta}}$; $x \in \mathbb{R}$; $\beta > 0$.

    V : $\frac{\pi^2}{6} \beta^2$.

    H : $\ln \beta + \gamma + 1$.

    Comment: γ is the Euler-Mascheroni constant.

12. **Inverse Chi-square** distribution.

    PDF: $f(x; \nu) = \frac{2^{-\nu/2}}{\Gamma(\nu/2)} x^{-\nu/2-1} e^{-1/(2x)}$; $x > 0$; $\nu > 0$.

    V : $\frac{2}{(\nu-2)^2(\nu-4)}$, $\nu > 4$.

    H : $\frac{\nu}{2} + \ln(\frac{1}{2}\Gamma(\frac{\nu}{2})) - (1 + \frac{\nu}{2})\psi(\frac{\nu}{2})$.

    Comment: a special case of the gamma distribution.

13. **Inverse Gaussian** distribution (also known as **Wald** distribution).

    PDF: $f(x; \lambda, \mu) = \left[\frac{\lambda}{2\pi x^3}\right]^{1/2} \exp \frac{-\lambda(x-\mu)^2}{2\mu^2 x}$; $x > 0$; λ, $\mu > 0$.

    V : μ^3/λ.

    H : $\frac{1}{2} + \frac{1}{2}\ln\frac{2\pi}{\lambda} + \frac{3\lambda}{\mu}e^{\frac{\lambda}{\mu}} - \frac{1}{\lambda} + 3e^{\frac{\lambda}{\mu}} \left(K_\delta \left(\frac{1}{\lambda}\right)\right)'_\delta \Big|_{\delta=-1/2}$.

    Comment: $H(V)$ is USF only for fixed m. The entropy [158] uses the derivative of the modified Bessel function of the second kind, which can be computed as $(K_\delta(x))'_\delta = -\frac{1}{2}(K_{\delta-1}(x) + K_{\delta+1}(x))$.

14. **Laplace** distribution.

    PDF: $f(x; \mu, \sigma) = \frac{1}{\sqrt{2}\sigma} \exp \left(-\frac{\sqrt{2}}{\sigma}|x - \mu|\right)$; $x \in \mathbb{R}$; $\sigma > 0$.

    V : σ^2.

    H : $\ln(\sigma e \sqrt{2})$.

15. **Logarithmic** distribution.

    PDF: $f(x; a, b) = K \ln x$ with $K = 1/[b(\ln b - 1) - a(\ln a - 1)]$; $x \in [a, b]$; $a, b > 0$, $a < b$.

    V : $\frac{K}{9}[b^3(3\ln b - 1) - a^3(3\ln a - 1)] - \mu^2$ with $\mu = \frac{K}{4}[b^2(2\ln b - 1) - a^2(2\ln a - 1)]$.

    H : $-K[b(\ln b - 2) - a(\ln a - 2) + Ei(\ln b) - Ei(\ln a)] - \ln K$, for $a, b > 1$.

    Comment: Ei is the exponential integral function.

16. **Logistic** distribution.

    PDF: $f(x; \mu, s) = \frac{e^{-(x-\mu)/s}}{s(1+e^{-(x-\mu)/s})^2}$; $x \in \mathbb{R}$; $s > 0$.

    V : $\frac{\pi^2}{3} s^2$.

    H : $\ln s + 2$.

17. **Log-Logistic** distribution.

    PDF: $f(x; \alpha, \beta) = \frac{(\beta/\alpha)(x/\alpha)^{\beta-1}}{[1+(x/\alpha)^\beta]^2}$; $x \geq 0$, α, $\beta > 0$.

    V : $\alpha^2 \left(\frac{2\pi/\beta}{\sin(2\pi/\beta)} - \frac{(\pi/\beta)^2}{\sin^2(\pi/\beta)}\right)$; $\beta > 2$.

    H : $2 - \ln \left(\frac{\alpha}{\beta}\right)$.

18. **Lognorm** distribution.

    PDF: $f(x; \mu, \sigma) = \frac{1}{x\sigma\sqrt{2\pi}}e^{-\frac{(\ln x - \mu)^2}{2\sigma^2}}$; $x > 0$, $\sigma > 0$.

    $V: (e^{\sigma^2} - 1)e^{2\mu+\sigma^2}$; $\beta > 2$.

    $H: \frac{1+\ln(2\pi\sigma^2)}{2} + \mu$.

19. **Maxwell-Boltzmann** distribution.

    PDF: $f(x; a) = \sqrt{\frac{2}{\pi}}\frac{x^2 e^{-x^2/(2a^2)}}{a^3}$; $x \geq 0$, $a > 0$.

    $V: \frac{a^2(3\pi-8)}{\pi}$.

    $H:$ (no known formula)

    Comment: a special case of the Gamma distribution.

20. **Pareto** distribution (also known as **Power law** distribution).

    PDF: $f(x; \alpha, x_m) = \frac{\alpha x_m^\alpha}{x^{\alpha+1}}$; $x > x_m$, α, $x_m > 0$.

    $V: \frac{\alpha x_m^2}{(\alpha-1)^2(\alpha-2)}$, $\alpha > 2$.

    $H: \ln\left(\frac{\alpha}{x_m}\right) - \frac{1}{\alpha} - 1$.

    Comment: a special case of Pearson Type XI distribution

21. **Pearson Type II** distribution.

    PDF: $f(x; a, m) = \frac{\Gamma(m+\frac{3}{2})}{a\sqrt{\pi}\Gamma(m+1)}\left(1 - \frac{x^2}{a^2}\right)^m$; $x \in [-a, a]$, $a > 0$, $m > -1$.

    $V: \frac{1}{2}\frac{a^2\Gamma(m+\frac{3}{2})}{\Gamma(m+\frac{5}{2})}$; $m > 5/2$.

    $H: -\ln\frac{\Gamma(m+\frac{3}{2})}{a\sqrt{\pi}\Gamma(m+1)} - \frac{4\Gamma(m+\frac{3}{2})ma^{1-2m}(\ln 2 - 1)}{a\sqrt{\pi}\Gamma(m+1)}$; $m > 3/2$.

    Comment: $H(V)$ is USF only for fixed m.

22. **Pearson Type IV** distribution.

    PDF: $f(x; k, a, m) = K(1 + (x/a)^2)^{-m}\exp(-k\text{atan}(x/a))$; $x \in \mathbb{R}$; a, $k > 0$, $m > 1/2$.

    $V: \frac{a^2}{r^2(r-1)}(r^2+1)$ for $k = 1$; $r = 2m - 2$, $m > 3/2$.

    $H: -\frac{I(r)}{F(r)} + \ln a + \ln F(r)$.

    with $F(r) = \int_{-\pi/2}^{\pi/2} e^{-\theta}\cos^r(\theta)d\theta$ and $I(r) = \int_{-\pi/2}^{\pi/2} e^{-\theta}\cos^r(\theta)((r + 2)\ln\cos\theta - \theta)d\theta$.

23. **Pearson Type V** distribution (also known as **Inverse-Gamma** distribution).

    PDF: $f(x; \alpha, \beta) = \beta^\alpha \frac{1}{x^{\alpha+1}}\exp(-\beta/x)/\Gamma(\alpha)$; $x > 0$; α, $\beta > 0$.

    $V: \beta^2/[(\alpha - 1)^2(\alpha - 2)]$.

    $H: \alpha + \ln(\beta\Gamma(\alpha)) - (1 + \alpha)\psi(\alpha)$.

24. **Pearson Type VII** distribution.

    PDF: $f(x; a, m) = K\left(1 + \frac{x^2}{a^2}\right)^{-m}$ with $K = a\sqrt{\pi}\frac{\Gamma(m-1/2)}{\Gamma(m)}$; $x \in \mathbb{R}$; $a \neq 0$, $m > 1/2$.

$V: \frac{a^2}{2}\frac{\Gamma(m-3/2)}{\Gamma(m-1/2)}$; $m > 3/2$.

$H: \frac{Kam}{(m-1)^2} - \ln\frac{K}{2}$; $m > 1$.

25. **Pearson Type VIII** distribution.

PDF: $f(x; a, m) = K(1 + x/a)^{-m}$ with $K = (m-1)/(\epsilon^{1-m} - 1)$;
$x \in [-a+\epsilon, 0]$, $\epsilon \in]0, a[$; $m > 1$.

$V: \frac{\epsilon(1-m)+(m-2)+\epsilon^{m-1}}{(m-2)(1-\epsilon^{m-1})} - \frac{\left((m-2)(\epsilon-1)+\epsilon-\epsilon^{m-1}\right)^2}{(m-2)^2(1-\epsilon^{m-1})^2}$; $m > 2$.

$H: \frac{m^2-m\epsilon^{m-1}}{(m-1)(1-\epsilon^{m-1})} - \ln\left(\frac{\epsilon^{1-m}-1}{m-1}\right)$; $m > 1$.

Comment: V and H computed for $a = 1$ (see D.2.1).

26. **Pearson Type IX** distribution.

PDF: $f(x; a, m) = K(1 + x/a)^m$ with $K = 1/(m+1)$; $x \in [-a, 0]$; $m > -1$.

$V: \frac{2}{m^2+5m+6} - \frac{1}{(m+2)^2}$.

$H: \frac{m}{m+1} - \ln(m+1)$.

Comment: V and H computed for $a = 1$ and $m > 0$ (see D.2.1).

27. **Pearson Type XI** distribution.

PDF: $f(x; m) = Kx^{-m}$ with $K = \epsilon^{1-m}/(m-1)$; $x \in [\epsilon, \infty[, \epsilon > 0; m > 1$.

$V: \frac{\epsilon^{2(m-2)}-2(m-1)\epsilon^{m-1}}{(m-2)^2} + \frac{m-1}{m-3}\epsilon^2$; $m > 3$.

$H: \frac{m}{m-1} + \ln(m-1) + (2m-1)\ln\epsilon$.

28. **Raised Cosine** distribution.

PDF: $f(x; s) = \frac{1}{2s}\left(1 + \cos\frac{\pi x}{s}\right)$; $x \in [-s, s]$, $s > 0$.

$V: s^2\left(\frac{1}{3} - \frac{2}{\pi^2}\right)$.

$H: 1 - 2\ln 2 - \ln(s)$.

29. **Rayleigh** distribution.

PDF: $f(x; \sigma) = \frac{x}{\sigma^2}e^{-x^2/2\sigma^2}$; $x \geq 0$.

$V: \frac{4-\pi}{2}\sigma^2$.

$H: 1 + \ln\frac{\sigma}{\sqrt{2}} + \gamma$.

Comment: γ is the Euler-Mascheroni constant.

30. **Student's t** distribution.

PDF: $f(x; \nu) = \frac{\Gamma\left(\frac{\nu+1}{2}\right)}{\sqrt{\nu\pi}\Gamma\left(\frac{\nu}{2}\right)}\left(1 + \frac{x^2}{\nu}\right)^{-\left(\frac{\nu+1}{2}\right)}$; $x \in \mathbb{R}$; $v > 0$.

$V: \frac{\nu}{\nu-2}$; $\nu > 2$.

$H: \frac{\nu+1}{2}\left[\psi\left(\frac{1+\nu}{2}\right) - \psi\left(\frac{\nu}{2}\right)\right] + \ln\left[\sqrt{\nu}B(\frac{\nu}{2}, \frac{1}{2})\right]$.

Comment: a special case of Pearson Type IV.

31. **Tanh-Neuron** distribution.

PDF: $f(x; a, b) = \frac{K}{1-x^2}\exp\left(-\frac{(\operatorname{atanh}(x)-a)^2}{b}\right)$, $x \in [-1, 1]$.

$V: $ (no known formula)

$H: $ (no known formula)

32. **Triangular** distribution.

   PDF: $f(x; w) = \begin{cases} -(w - x\ \text{sgn}(x))/w^2 & , \ x \in [-w, w], \ w > 0 \\ 0 & , \ \text{otherwise} \end{cases}$.

   $V : \frac{w^2}{6}$.

   $H : \frac{1}{2} + \ln w$.

33. **Uniform** distribution.

   PDF: $f(x; a, b) = \begin{cases} 1/(b - a) & , \ x \in [a, b] \\ 0 & , \ \text{otherwise} \end{cases}$.

   $V : \frac{(b-a)^2}{12}$.

   $H : \ln(b - a)$.

   Comment: A special case of Pearson Type II.

34. **Weibull** distribution.

   PDF: $f(x; \lambda, \ k) = \frac{k}{\lambda} \left(\frac{x}{\lambda}\right)^{k-1} e^{-(x/\lambda)^k}$; $x \geq 0$; $\lambda, \ k > 0$.

   $V : \lambda^2 \Gamma \left(1 + \frac{2}{k}\right) - \lambda^2 \Gamma^2 \left(1 + \frac{1}{k}\right)$.

   $H : \gamma \left(1 - \frac{1}{k}\right) + \ln \left(\frac{\lambda}{k}\right) + 1$.

   Comment: γ is the Euler-Mascheroni constant.

Appendix E
Optimal Parzen Window Estimation

E.1 Parzen Window Estimation

Let us consider the problem of estimating a univariate PDF $f(x)$ of a continuous r.v. X based on a i.i.d. sample $X_n = \{x_1, \ldots, x_n\}$. Although unbiased estimators do not exist in general for $f(x)$ (see e.g., [181]), it is possible nonetheless to define sequences of density estimators, $\hat{f}_n(x) \equiv \hat{f}_n(x; X_n)$, *asymptotically unbiased*:

$$\lim_{n\to\infty} \mathbb{E}_X[\hat{f}_n(x)] = f(x) . \tag{E.1}$$

The Parzen window estimator [169] provides such an asymptotically unbiased estimate. It is a generalization of the shifted-histogram estimator (see e.g., [224]), defined as:

$$\hat{f}_n(x) = \frac{1}{n}\sum_{i=1}^{n} \frac{1}{h} K\left(\frac{x - x_i}{h}\right) , \tag{E.2}$$

where the positive constant $h \equiv h(n)$ is the so-called *bandwidth* of $K(x)$, the *Parzen window* or *kernel function*, which is any Lebesgue measurable function satisfying:

i Boundedness: $\sup_{\mathbb{R}}|K(x)| < \infty$;

ii $K(x)$ belongs to the $\mathcal{L}^1$ space, i.e., $\int |K(x)| < \infty$;
iii $K(x)$ decreases faster than $1/x$: $\lim_{x\to\infty} |xK(x)| = 0$;
iv $\int K(x) = 1$.

For reasons to be presented shortly, the Gaussian kernel, $G(x) = \exp(-x^2/2)$ $/\sqrt{2\pi}$, is a popular choice for kernel function $K(x)$.
Sometimes the Parzen window estimator is written as

$$\hat{f}_n(x) = \frac{1}{n}\sum_{i=1}^{n} \frac{1}{h} K\left(\frac{x - X_i}{h}\right) , \tag{E.3}$$

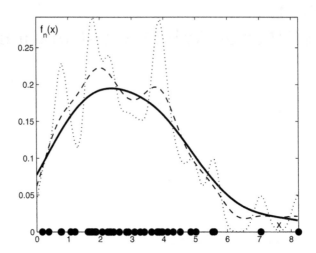

Fig. E.1 Parzen window PDF estimates for the data instances represented by the black circles. The estimates are obtained with a Gaussian kernel with bandwidths: $h = 0.2$ (dotted line), $h = 0.5$ (dashed line), and $h = 1$ (solid line).

to emphasize the fact that $\hat{f}_n(x)$ is a random variable with a distribution dependent on the joint distribution of the i.i.d. X_i.

The Parzen window estimator can also be written as a convolution of the Parzen window with the empirical distribution:

$$\hat{f}_n(x) = K_h \otimes \mu_n(x) = \int \frac{1}{h} K\left(\frac{x-y}{h}\right) \mu_n(y) dy , \qquad (E.4)$$

where $\mu_n(x) = \sum_{i=1}^{n} \delta(x - x_i)$ is a Dirac-δ comb representing the empirical density, and $K_h(x) = \frac{1}{h} K\left(\frac{x}{h}\right)$. We may also write $K_h(x)$ as $K(x; h)$; $K(x)$ is then $K(x; 1)$. Note that $\int |K_h(x)| = \int |K(x)|$.

For kernels satisfying the above conditions the convolution operation yields an estimate $\hat{f}_n(x)$ which is a smoothed version of $f(x)$. The degree of smoothing increases with h, as exemplified by Fig. E.1, showing three different estimates of a PDF computed with a Gaussian kernel on a 40 instance dataset, random and independently drawn from the chi-square distribution with three degrees of freedom.

The Parzen window estimate $\hat{f}_n$ enjoys the following important properties:

1. For a dataset with sample mean $\bar{x}$ and sample variance s^2, if K is a symmetric function, the mean μ_n and the variance σ_n^2 of $\hat{f}_n$ satisfy:

$$\mu_n = \bar{x}; \qquad\qquad \sigma_n^2 \equiv V[\hat{f}_n(x)] = s^2 + h^2 \int x^2 K(x) dx , \qquad (E.5)$$

2. If $h \underset{n\to\infty}{\to} 0$ with $nh \underset{n\to\infty}{\to} \infty$ (h decreases less than $1/n$) the $\hat{f}_n$ estimate verifies:

$$\lim_{n\to\infty} nhV[\hat{f}_n(x)] = f(x) \int K^2(y)dy .\tag{E.6}$$

For the Gaussian kernel: $\lim_{n\to\infty} V[\hat{f}_n(x)] = \frac{f(x)}{2nh\sqrt{\pi}}$.

3. If K satisfies the conditions stated in 1 and 2, the estimate $\hat{f}_n$ is then MSE-consistent: $MSE(\hat{f}_n(x))_{n\to\infty} \to 0$ (with $MSE(\hat{f}_n(x)) = \mathbb{E}[(\hat{f}_n(x) - f(x))^2]$).

4. The MSE-consistent $\hat{f}_n$, for a density having r derivatives, verifies:

$$MSE(\hat{f}_n(x)) \cong \frac{f(x)}{nh} \int_{-\infty}^{\infty} K^2(y)dy + h^{2r}k_r^2|f^{(r)}(x)|^2 ,\tag{E.7}$$

where $f^{(r)}(x)$ is the derivative of order r and k_r is the *characteristic exponent* of the Fourier transform of $K(x)$ (denoted $k(u)$) defined as:

$$k_r = \lim_{u\to 0} \frac{1 - k(u)}{|u|^r} .\tag{E.8}$$

Any symmetric kernel such that $x^2 K(x) \in \mathcal{L}^1$, has a nonzero finite k_r for $r = 2$. In particular, for the Gaussian kernel $k_r = 1/2$ for $r = 2$.

5. The optimal (smallest) MSE of the MSE-consistent $\hat{f}_n$, has the following value at a given $x \in X$:

$$MSE_{opt}(\hat{f}_n(x)) \cong$$
$$(2r + 1)\left\{ \frac{f(x)}{2nr} \int_{-\infty}^{\infty} K^2(y)dy \right\}^{2r/(2r+1)} \left| k_r f^{(r)}(x) \right|^{2r/(2r+1)} .\tag{E.9}$$

Thus, the decrease of the optimal MSE with the sample size n is of order $n^{-2r/(2r+1)}$.

6. The optimal IMSE (the integrated MSE for the whole X support) of the MSE-consistent $\hat{f}_n$, for a symmetric kernel such that $x^2 K(x) \in \mathcal{L}^1$, is obtained with a bandwidth given by:

$$h(n) = n^{-1/5}\alpha(K)\beta(f) ,\tag{E.10}$$

with

$$\alpha(K) = \left[\frac{\int K^2(y)dy}{(\int y^2 K(y)d_y)^2} \right]^{1/5} \quad \text{and} \quad \beta(f) = \left[\int \left| f^{(2)}(y) \right|^2 dy \right]^{-1/5} .\tag{E.11}$$

For the Gaussian kernel $\alpha(K) = 0.7764$. The $\beta(f)$ factor is usually unknown. One could consider iteratively improving an initial estimate of $\beta(f)$ as discussed in [224].

7. The optimal IMSE in the conditions of 6, attains its smallest value for the Epanechnikov kernel defined as:

$$K_e(t) = \begin{cases} \frac{3}{4\sqrt{5}}\left(1 - \frac{1}{5}t^2\right), & -\sqrt{5} \le t \le \sqrt{5} \\ 0 & , \text{ otherwise} \end{cases} . \qquad (E.12)$$

The monograph [221] indicates the efficiencies (optimal IMSE ratios) of other kernels compared to K_e. The efficiency of the Gaussian kernel is ≈ 0.9512.

8. The deviations (residues) of the MSE-consistent $\hat{f}_n$ are asymptotically normal:

$$\lim_{n \to \infty} P\left\{ \frac{\hat{f}_n(x) - \mathbb{E}[\hat{f}_n(x)]}{\sigma[\hat{f}_n(x)]} \le c \right\} = g(c; 0, 1) . \qquad (E.13)$$

9. By the bounded difference inequality the following result holds:

$$P\left(\left| \int |f_n - f| - E\left[\int |f_n - f| \right] \right| \ge t \right) \le 2e^{-t^2/2n(\int |K|)^2} . \qquad (E.14)$$

10. Parzen window estimates are stable: suppose that one of the X_i changes value while the other $n-1$ data instances remain fixed; let $\hat{f}_n^*$ denote the new perturbed estimate; then: $\int |\hat{f}_n - \hat{f}_n^*| \le \frac{2}{n} \int |K|$.

The above properties are discussed in detail in the following references: properties 1 through 7, in [224]; property 8 in [235]; properties 9 and 10 in [53].

The efficiency of the Gaussian kernel mentioned in property 7, together with the ease of implementation and the ease of analysis resulting from the remarks mentioned in properties 2, 4, and 6, justify its popularity.

E.2   Optimal Bandwidth and IMSE

The choice of a kernel bandwidth adequate to the sample size n is of the utmost importance in accurate Parzen window estimation of PDFs. For this purpose, and assuming the kernel function to satisfy the conditions mentioned in property 6, we apply formulas (E.10) and (E.11) to the computation of the optimal bandwidth. The corresponding IMSE formula for $r = 2$, is [224]:

$$IMSE \cong \frac{1}{nh} \int_{-\infty}^{\infty} K^2(y)dy + \frac{1}{4}h^4 \int_{-\infty}^{\infty} \left|f''(x)\right|^2 dx . \qquad (E.15)$$

Table E.1 Optimal h and $IMSE$ (per n) for the Gamma distribution with $b = 1$

a	I_2	h	$IMSE$	$\sigma = b\sqrt{a}$
1	0.5000	0.8918	0.7908	1.0000
2	1.2500	0.7425	0.9498	1.4142
3	0.1875	1.0851	0.6499	1.7321
4	0.0313	1.5528	0.4542	2.0000

We now proceed to simplify these formulas, using the following simplified notation: α, β instead of $\alpha(K), \beta(f)$; $I_K = \int K^2$; $I_2 = \int |f''|^2$. The optimal h and IMSE can then be written as:

$$h = \left(\frac{I_K}{I_2}\right)^{0.2} n^{-0.2}; \quad IMSE = \frac{I_K}{nh} + \frac{h^4 I_2}{4} = \frac{5h^4}{4} I_2 . \tag{E.16}$$

For the Gaussian kernel we have:

$$h = \left(\frac{0.282095}{I_2}\right)^{0.2} n^{-0.2}; \quad IMSE = 0.454178 I_2^{0.2} n^{-0.8} . \tag{E.17}$$

Thus, in order to compute the optimal bandwidth and IMSE when using a Gaussian kernel, the only thing that remains to be done is to compute I_2. For instance, for the Gaussian density with standard deviation σ, we compute:

$$\int |f''(y)|^2 dy \approx 0.212\sigma^{-5} \Rightarrow \beta(f) \approx 1.3637\sigma . \tag{E.18}$$

Hence:

$$h = 1.0592\sigma n^{-0.2}; \quad IMSE = \frac{0.332911}{\sigma} n^{-0.8} . \tag{E.19}$$

The difficulty with formulas (E.17) is the need to compute I_2 dependent on the second derivative of $f(x)$ which may be unknown. For symmetrical PDFs, reasonably close to the Gaussian PDF, formulas (E.19) are satisfactory. For other distributions one has to compute the integral I_2. As an example, Table E.1 shows the optimal h and $IMSE$, per n, for a few cases of the gamma distribution:

$$\gamma(x; a, b) = \frac{1}{b^a \Gamma(a)} x^{a-1} e^{-x/b} \text{ for } x > 0, \text{ with } a > 0 \text{ (shape)}, b > 0 \text{ (scale)} . \tag{E.20}$$

Note that $a = 1$ corresponds to the special exponential case of the gamma distribution.

Let us consider the $a = 2$ case. For equal σ one obtains for the normal distribution, per n, $h = 1.4979$ (with $IMSE = 0.2354$). The lower h for the gamma distribution reflects its skewed shape, with a far steeper left tail than the normal distribution; the $IMSE$ is larger, though, since the decreased h doesn't account well for the longer right tail. Similar observations can be drawn from other cases.

Whenever the value of I_2 cannot be obtained, the empirically adjusted formula $h = 0.79Rn^{-1/5}$, where R is the interquartile range [221], may produce satisfying results.

Appendix F
Entropy Estimation

Let us assume an i.i.d. sample $X_n = (x_1, \ldots, x_n)$ drawn from some univariate continuous distribution with unknown PDF $f(x)$. The general problem to be addressed is how to use X_n in order to obtain an estimate of an $f(x)$ functional, such as entropy, $H(X)$. We are here particularly interested in estimating the Shannon's and Rényi's quadratic entropies.

Estimators of PDF functionals can be of four types [23]: integral estimator; plug-in estimator; splitting data estimator; cross-validation estimator.

F.1 Integral and Plug-in Estimates

The integral estimator corresponds to an idea that immediately jumps to mind: substitute $f(x)$ by an estimate $\hat{f}_n(x)$ in the formula of the functional. When the functional is the Shannon entropy, this amounts to computing:

$$\hat{H}_S(X) = - \int \hat{f}_n(x) \ln \hat{f}_n(x) dx \ . \tag{F.1}$$

Unfortunately, in this case the computation of $\hat{H}_S(x)$ requires numerical integration. As an alternative for such cases, one can substitute $f(x)$ by $\hat{f}_n(x)$ in the *empirical expression* of the functional. This corresponds to the plug-in (or resubstitution) estimator and is usually easily implemented.

F.2 Integral Estimate of Rényi's Quadratic Entropy

The integral estimate of Rényi's quadratic entropy has a computationally interesting form, that doesn't raise the above mentioned problem of the need of numerical integration, when the PDF estimate $\hat{f}_n(x)$ is obtained by the

Parzen window method with the Gaussian kernel [246, 245]. The estimate $\hat{f}_n(x)$ is then written as (see Appendix E):

$$\hat{f}_n(x) = \frac{1}{nh} \sum_{i=1}^{n} G\left(\frac{x - x_i}{h}\right) = \frac{1}{n} \sum_{i=1}^{n} G_h(x - x_i) , \qquad (F.2)$$

where $G_h(x)$ is the Gaussian kernel of bandwidth h (same role as the standard deviation of the Gaussian PDF)

$$G_h(x) = \frac{1}{\sqrt{2\pi}h} \exp\left(-\frac{x^2}{2h^2}\right) . \qquad (F.3)$$

Substituting the estimate $\hat{f}_n(x)$ in the formula of $H_{R_2}(x)$, one obtains the integral estimate:

$$\hat{H}_{R_2}(X) = -\ln \int_{-\infty}^{+\infty} \left(\frac{1}{n} \sum_{i=1}^{n} G_h(x - x_i)\right)^2 dx =$$

$$= -\ln \left[\frac{1}{n^2} \int_{-\infty}^{+\infty} \left(\sum_{i=1}^{n} G_h(x - x_i)\right)^2 dx\right] . \qquad (F.4)$$

Since we have a finite sum we may interchange sum and integration and write this integral estimate as:

$$\hat{H}_{R_2}(X) = -\ln \left[\frac{1}{n^2} \sum_{i=1}^{n} \sum_{j=1}^{n} \int_{-\infty}^{+\infty} G_h(x - x_i) G_h(x - x_j) dx\right] . \qquad (F.5)$$

We now use the following theorem (a stronger version is proved in [245]):

Theorem F.1. *Let $g(x; a, \sigma)$ and $g(x; b, \sigma)$ be two Gaussian functions with equal variance. The integral of their product is a Gaussian whose mean is the difference of the means, $a - b$, and whose variance is the double of the original variance, $2\sigma^2$.*

Proof. We have

$$g(x; a, \sigma)g(x; b, \sigma) = \frac{1}{2\pi\sigma^2} \exp\left[-\frac{1}{2}\frac{(x - a)^2 + (x - b)^2}{\sigma^2}\right] . \qquad (F.6)$$

By adding and subtracting $(a + b)^2/2$ to the numerator of the exponent, we express it as $2(x - (a + b)/2)^2 + (a - b)^2/2$. Therefore:

$$g(x; a, \sigma)g(x; b, \sigma) = \frac{1}{\sqrt{2\pi}\frac{\sigma}{\sqrt{2}}} \exp\left[-\frac{1}{2}\frac{(x - (a + b)/2)^2}{(\sigma/\sqrt{2})^2}\right]$$

$$\frac{1}{\sqrt{2\pi}(\sqrt{2}\sigma)} \exp\left[-\frac{1}{2}\frac{(a - b)^2}{(\sqrt{2}\sigma)^2}\right] =$$

$$= g\left(x; \frac{a + b}{2}, \frac{\sigma}{\sqrt{2}}\right) g(0; a - b, \sqrt{2}\sigma) . \qquad (F.7)$$

Since the integral of a Gaussian function is 1, we then obtain:

$$\int_{-\infty}^{+\infty} g(x; a, \sigma)g(x; b, \sigma)dx = g(0; a - b, \sqrt{2}\sigma) . \qquad (F.8)$$

□

Applying this theorem to the integral in (F.5), we get (using the kernel notation)

$$\hat{H}_{R_2}(X) = -\ln\left[\frac{1}{n^2}\sum_{i=1}^{n}\sum_{j=1}^{n}G_{\sqrt{2}h}(x_i - x_j)\right] . \qquad (F.9)$$

We then obtain an estimate of Rényi's quadratic entropy directly computable in terms of Gaussian functions.

Note that the MSE consistency of the Parzen window estimate (see Appendix E) directly implies the MSE consistency of this $\hat{H}_{R_2}$ estimate.

F.3 Plug-in Estimate of Shannon's Entropy

When H is the Shannon entropy, $H_S(x) = -\int f(x)\ln f(x)dx$, which is the expected value of $\ln f(x)$, we plug-in the PDF estimate in the empirical formula of the expectation and obtain [1]:

$$\hat{H}_S(X) = -\frac{1}{n}\sum_{i=1}^{n}\ln \hat{f}_n(x_i) . \qquad (F.10)$$

When $\hat{f}_n(x)$ is obtained by the Parzen window method with kernel K and bandwidth h, we have:

$$\hat{H}_S(X) = -\frac{1}{n}\sum_{i=1}^{n}\ln\sum_{j=1}^{n}K_h(x_i - x_j) . \qquad (F.11)$$

The $\hat{H}_S(X)$ estimate enjoys the following consistency properties [1, 159]:

- $\mathscr{L}_1$ consistency: if $nh \to \infty$ as $n \to \infty$, $\int [\ln f(x)]^2 f(x)dx < \infty$, $f'(x)$ is continuous and $\sup |f'(x)| < \infty$, and $\int |u|K(u)du < \infty$, then $\mathbb{E}[\|\hat{H}_S(X) - H_S(X)\|] \underset{n\to\infty}{\to} 0$.
- $\mathscr{L}_2$ consistency: if, in addition, $\int (f'(x)/f(x))^2 f(x)dx < \infty$ (finite Fisher information number), then $\mathbb{E}[\|\hat{H}_S(X) - H_S(X)\|^2] \underset{n\to\infty}{\to} 0$.
- MSE consistency, a consequence of the Parzen window MSE consistency.
- Almost sure (a.s.) consistency: $\hat{H}_S(X) \underset{n\to\infty}{\to} H_S(X)$ a.s., under certain mild conditions (see [159]).

F.4 Plug-in Estimate of Rényi's Entropy

Let us consider the expression of Rényi's entropy of order α:

$$H_{R_\alpha}(X) = \frac{1}{1-\alpha} \ln \int_E f^\alpha de = \frac{1}{1-\alpha} \ln V_\alpha, \quad \alpha \geq 0 . \tag{F.12}$$

Since the information potential V_α is the mean of $f^{\alpha-1}(x)$, the plug-in estimator is immediately written as:

$$\hat{H}_{R_\alpha}(X) = \frac{1}{1-\alpha} \ln \left(\frac{1}{n} \sum_{i=1}^n \hat{f}_n^{\alpha-1}(x_i) \right) =$$

$$= \frac{1}{1-\alpha} \ln \left(\frac{1}{n^\alpha} \sum_{i=1}^n \sum_{j=1}^n K_h(x_i - x_j) \right) . \tag{F.13}$$

The MSE consistency of the Parzen window estimate (see Appendix E) directly implies the MSE consistency of the $\hat{H}_{R_\alpha}$ estimate. Note that for the quadratic entropy with Gaussian kernel there is a difference of $\sqrt{2}$ in the bandwidths of both estimates (F.9) and (F.13). This difference is unimportant in practical terms.

References

1. Ahmad, I.A., Lin, P.: A Nonparametric Estimation of the Entropy for Absolutely Continuous Distributions. IEEE Trans. Information Theory 22(3), 372–375 (1976)
2. Alexandre, L.A.: Maximizing the zero-error density for RTRL. In: 8th IEEE Int. Symposium on Signal Processing and Information Technology, IEEE Press, Sarajevo (2008)
3. Alexandre, L.A.: Single Layer Complex Valued Neural Network with Entropic Cost Function. In: Honkela, T. (ed.) ICANN 2011, Part I. LNCS, vol. 6791, pp. 331–338. Springer, Heidelberg (2011)
4. Alexandre, L.A., Campilho, A., Kamel, M.: A Probabilistic Model for the Cooperative Modular Neural Network. In: Perales, F.J., Campilho, A.C., Pérez, N., Sanfeliu, A. (eds.) IbPRIA 2003. LNCS, vol. 2652, pp. 11–18. Springer, Heidelberg (2003)
5. Alexandre, L.A., Campilho, A., Kamel, M.: Bounds for the Average Generalization Error of the Mixture of Experts Neural Network. In: Fred, A., Caelli, T.M., Duin, R.P.W., Campilho, A.C., de Ridder, D. (eds.) SSPR&SPR 2004. LNCS, vol. 3138, pp. 618–625. Springer, Heidelberg (2004)
6. Alexandre, L.A., Marques de Sá, J.: Error Entropy Minimization for LSTM Training. In: Kollias, S.D., Stafylopatis, A., Duch, W., Oja, E. (eds.) ICANN 2006. LNCS, vol. 4131, pp. 244–253. Springer, Heidelberg (2006)
7. Alexandre, L.A., Silva, L., Santos, J., Marques de Sá, J.: Experimental evaluation of multilayer perceptrons with entropic risk functionals on real-world datasets. Technical Report 2/2011, INEB - Instituto de Engenharia Biomédica, Porto, Portugal (2011)
8. Allwein, E., Schapire, R., Singer, Y.: Reducing multiclass to binary: A unifying approach for margin classifiers. In: Int. Conf. on Machine Learning, pp. 9–16. Morgan Kaufmann (2000)
9. Amin, M., Murase, K.: Single-layered complex-valued neural network for real-valued classification problems. Neurocomputing 72, 945–955 (2009)
10. Andreev, A., Kanto, A., Malo, P.: Simple Approach for Distribution Selection in the Pearson System, Working Paper W388, Helsinki School of Economics (2005)
11. Anthony, M., Bartlett, P.: Neural network learning: theoretical foundations. Cambridge University Press (1999)
12. Argentiero, P., Chin, R., Beaudet, P.: An automated approach to the design of decision tree classifiers. IEEE Trans. on Pattern Analysis and Machine Intelligence 4(1), 51–57 (1982)

13. Asuncion, A., Newman, D.: UCI machine learning repository, University of California, School of Information and Computer Science (2010), http://www.ics.uci.edu/~mlearn/MLRepository.html
14. Auda, G., Kamel, M.: Modular neural network classifiers: A comparative study. Intel. Robotic Systems 21, 117–129 (1998)
15. Balakrishnan, N., Nevzorov, V.: A Primer on Statistical Distributions. John Wiley & Sons, Inc. (2003)
16. Banarer, V., Perwass, C., Sommer, G.: Design of a Multilayered Feed-Forward Neural Network Using Hypersphere Neurons. In: Petkov, N., Westenberg, M.A. (eds.) CAIP 2003. LNCS, vol. 2756, pp. 571–578. Springer, Heidelberg (2003)
17. Banarer, V., Perwass, C., Sommer, G.: The hypersphere neuron. In: European Symposium on Artificial Neural Networks, pp. 469–474 (2003)
18. Batchelor, B.: Practical Approach to Pattern Classification. Plenum Press, Plenum Pub. Co. Ltd. (1974)
19. Batchelor, B.: Classification and data analysis in vector spaces. In: Batchelor, B. (ed.) Pattern Recognition: Ideas in Practice, Plenum Press, Plenum Pub. Co. Ltd. (1978)
20. Battiti, R.: Accelerated backpropagation learning: Two optimization methods. Complex Systems 3, 331–342 (1989)
21. Bauer, E., Kohavi, R.: An empirical comparison of voting classification algorithms: Bagging, boosting, and variants. Machine Learning 36(1-2), 105–139 (1999)
22. Baum, E., Wilczek, F.: Supervised learning of probability distributions by neural networks. In: NIPS 1987, pp. 52–61 (1987)
23. Beirlant, J., Dudewicz, E.J., Györfi, L., van der Meulen, E.C.: Nonparametric Entropy Estimation: An Overview. Int. J. Math. Stat. Sci. 6(1), 17–39 (1997)
24. Beiu, V., Pauw, T.: Tight Bounds on the Size of Neural Networks for Classification Problems. In: Cabestany, J., Mira, J., Moreno-Díaz, R. (eds.) IWANN 1997. LNCS, vol. 1240, pp. 743–752. Springer, Heidelberg (1997)
25. Berkhin, P.: Survey of clustering data mining techniques. Tech. rep., Accrue Software, San Jose, CA (2002)
26. Bishop, C.: Neural Networks for Pattern Recognition. Oxford University Press (1995)
27. Bishop, C.: Pattern Recognition and Machine Learning, 2nd edn. Springer (2007)
28. Blakrishnan, N., Navzorov, V.B.: A Primer on Statistical Distributions. John Wiley & Sons (2003)
29. Blum, A., Rivest, R.: Training a 3-node Neural Network is NP-Complete. In: Hanson, S.J., Rivest, R.L., Remmele, W. (eds.) MIT-Siemens 1993. LNCS, vol. 661, pp. 9–28. Springer, Heidelberg (1993)
30. Bousquet, O., Boucheron, S., Lugosi, G.: Introduction to statistical learning theory. In: Bousquet, O., von Luxburg, U., Rätsch, G. (eds.) Machine Learning 2003. LNCS (LNAI), vol. 3176, pp. 169–207. Springer, Heidelberg (2004)
31. Bowman, A., Azzalini, A.: Applied Smooting Techniques for Data Analysis. Oxford University Press (1997)
32. Brady, M., Raghavan, R., Slawny, J.: Gradient descent fails to separate. In: IEEE Int. Conf. on Neural Networks, vol. 1, pp. 649–656 (1988)
33. Breiman, L., Friedman, J., Olshen, R., Stone, C.: Classification and Regression Trees. Chapman & Hall/CRC (1993)
34. Buntine, W.: A theory of learning classification rules. Ph.D. thesis, Univ. of Technology, Sidney (1990)
35. Buntine, W., Niblett, T.: A further comparison of splitting rules for decision-tree induction. Machine Learning 8, 75–85 (1992)

36. Chan, L., Fallside, F.: An adaptive training algorithm for backpropagation networks. Computer Speech and Language 2, 205–218 (1987)
37. Chen, B., Hu, J., Zhu, Y., Sun, Z.: Information theoretic interpretation of error criteria. Acta Automatica Sinica 35(10), 1302–1309 (2009)
38. Chen, T., Geman, S.: On the minimum entropy of a mixture of unimodal and symmetric distributions. IEEE Trans. Information Theory 54(7), 1366–1374 (2008)
39. Cheng, C., Fu, A., Zhang, Y.: Entropy-based subspace clustering for mining numerical data. In: Int. Conf. on Knowledge Discovery and Data Mining, pp. 84–93 (1999)
40. Cheng, Y.: Mean shift, mode seeking, and clustering. IEEE Trans. on Pattern Analysis and Machine Intelligence 17(8), 790–799 (1995)
41. Cherkassky, V., Mulier, F.: Learning from data: concepts, theory and methods. John Wiley & Sons (1998)
42. Chung, F.: Spectral Graph Theory, vol. 92. American Mathematical Society, Providence (1997)
43. Collobert, R., Bengio, S.: Links between perceptrons, MLPs and SVMs. In: Proc. of the 21st Int. Conf. on Machine Learning (2004)
44. Comaniciu, D., Meer, P.: Mean shift analysis and applications. In: IEEE Int. Conf. on Computer Vision, pp. 1197–1203 (1999)
45. Comaniciu, D., Meer, P.: Mean shift: A robust approach toward feature space analisys. IEEE Trans. on Pattern Analysis and Machine Intelligence 24(5), 603–619 (2002)
46. Cortes, C., Vapnik, V.: Support-vector networks. Machine Learning 20(3), 273–297 (1995)
47. Costa, J., Hero, A., Vignat, C.: On Solutions to Multivariate Maximum alpha-Entropy Problems. In: Rangarajan, A., Figueiredo, M., Zerubia, J. (eds.) EMMCVPR 2003. LNCS, vol. 2683, pp. 211–226. Springer, Heidelberg (2003)
48. Cover, T.M., Thomas, J.A.: Elements of Information Theory. Wiley Interscience (1991)
49. Deco, G., Obradovic, D.: An Information-Theoretic Approach to Neural Computing. Springer (1996)
50. Dembo, A., Cover, T.M., Thomas, J.A.: Information theoretic inequalities. IEEE Trans. on Information Theory 37(6), 1501–1518 (1991)
51. Demšar, J.: Statistical comparisons of classifiers over multiple data sets. J. of Machine Learning Research 7, 1–30 (2006)
52. Devroye, L., Györfi, L., Lugosi, G.: A Probabilistic Theory of Pattern Recognition. Springer (1996)
53. Devroye, L., Lugosi, G.: Combinatorial Methods in Density Estimation. Springer (2001)
54. Dietterich, T.: An experimental comparison of three methods for constructing ensembles of decision trees: Bagging, boosting, and randomization. Machine Learning 40(2), 139–157 (2000)
55. Ding, C., He, X., Zha, H., Gu, M., Simon, H.: A min-max cut algorithm for graph partitioning and data clustering. In: Int. Conf. on Data Mining, pp. 107–114 (2001)
56. Dixon, W., Massey, F.: Introduction to Statistical Analysis, 4th edn. McGraw-Hill Companies (1983)
57. Draghici, S., Beiu, V.: Entropy based comparison of neural networks for classification. In: Italian Conf. on Neural Networks (1997)
58. Du, J.X., Zhai, C.M.: Structure optimization algorithm for radial basis probabilistic neural networks based on the moving median center hyperspheres algorithm. Advances in Neural Networks 55, 136–143 (2009)
59. Duda, R., Hart, P., Stork, D.: Pattern Classification. Wiley Interscience (2001)

60. Duin, R.: Dutch handwritten numerals,
    http://www.ph.tn.tudelft.nl/~duin
61. Dwass, M.: Probability: Theory and Applications. W. A. Benjamin, Inc. (1970)
62. Ebrahimi, N., Soofi, E., Soyer, R.: Information measures in perspective. Int.
    Statistical Review 78, 383–412 (2010)
63. Elman, J.: Finding structure in time. Tech. Rep. CRL 8801, University of
    California, Center for Research in Language, San Diego (1988)
64. Ennaji, A., Ribert, A., Lecourtier, Y.: From data topology to a modular clas-
    sifier. Int. Journal on Document Analysis and Recognition 6(1), 1–9 (2003)
65. Erdogmus, D., Hild II, K., Principe, J.: Blind source separation using Rényi's
    α–marginal entropies. Neurocomputing 49, 25–38 (2002)
66. Erdogmus, D., Principe, J.: Comparison of entropy and mean square error
    criteria in adaptive system training using higher order statistics. In: Int. Conf.
    on ICA and Signal Separation, Helsinki, Finland, pp. 75–80 (2000)
67. Erdogmus, D., Principe, J.: An error-entropy minimization algorithm for su-
    pervised training of nonlinear adaptive systems. IEEE Trans. on Signal Pro-
    cessing 50(7), 1780–1786 (2002)
68. Esposito, F., Malerba, D., Semeraro, G., Kay, J.: A comparative analysis of
    methods for pruning decision trees. IEEE Trans. on Pattern Analysis and
    Machine Intelligence 19(5), 476–493 (1997)
69. Ester, M., Kriegel, H.P., Sander, J., Xu, X.: A density-based algorithm for
    discovering clusters in large spatial databases with noise. In: 2nd Int. Conf.
    on Knowledge Discovery and Data Mining, pp. 226–231. AAAI Press (1996)
70. Álvarez Estévez, D., Príncipe, J., Moret-Bonillo, V.: Neuro-fuzzy classification
    using the correntropy criterion: Application to sleep depth estimation. In:
    Arabnia, H., de la Fuente, D., Kozerenko, E., Olivas, J., Chang, R., LaMonica,
    P., Liuzzi, R., Solo, A. (eds.) IC-AI, pp. 9–15. CSREA Press (2010)
71. Faivishevsky, L., Goldberger, J.: A nonparametric information theoretic clus-
    tering algorithm. In: Proc. of the 27th Int. Conf. on Machine Learning, pp.
    351–358 (2010)
72. Fhalman, S.: Faster-learning variations on back-propagation: An empirical
    study. In: Touretzky, D., Hinton, G., Sejnowski, T. (eds.) Connectionist Mod-
    els Summer School, pp. 38–51. Morgan Kaufmann (1988)
73. Fiedler, M.: A property of eigenvectors of nonnegative symmetric matrices and
    its application to graph theory. Czechoslovak Mathematical Journal 25(100),
    619–633 (1975)
74. Foley, D.: Considerations of sample and feature size. IEEE Trans. Information
    Theory 18(5), 618–626 (1972)
75. Forina, M., Armanino, C.: Eigenvector projection and simplified non-linear
    mapping of fatty acid content of italian olive oils. Ann. Chim. 72, 127–155
    (1981)
76. Fukunaga, K.: Introduction to Statistical Pattern Recognition, 2nd edn. Aca-
    demic Press Professional, Inc. (1990)
77. Fukunaga, K., Hostetler, L.: The estimation of the gradient of a density func-
    tion, with applications in pattern recognition. IEEE Trans. in Information
    Theory 21, 32–40 (1975)
78. Funahashi, K.: Multilayer neural networks and Bayes decision theory. Neural
    Networks 11(2), 209–213 (1998)
79. García, S., Fernández, A., Luengo, J., Herrera, F.: Advanced nonparametric
    tests for multiple comparisons in the design of experiments in computational
    intelligence and data mining: Experimental analysis of power. Information
    Sciences 180, 2044–2064 (2010)
80. Gelfand, S., Ravishankar, C., Delp, E.: An iterative growing and pruning al-
    gorithm for classification tree design. IEEE Trans. on Pattern Analysis and
    Machine Intelligence 13(2), 163–174 (1991)

81. Gers, F., Schmidhuber, J.: Recurrent nets that time and count. In: Int. Joint Conf. on Neural Networks, Como, Italy, pp. 189–194 (2000)

82. Gers, F., Schmidhuber, J., Cummins, F.: Learning to forget: Continual prediction with LSTM. Neural Computation 12(10), 2451–2471 (2000)

83. Gish, H.: A probabilistic approach to the understanding and training of neural network classifiers. In: Int. Conf. on Acoustics, Speech, and Signal Processing, vol. 3, pp. 1361–1364 (1990)

84. Gokcay, E., Principe, J.: Information theoretic clustering. IEEE Trans. on Pattern Analysis and Machine Intelligence 24(2), 158–171 (2002)

85. Guha, S., Rastogi, R., Shim, K.: CURE: an efficient clustering algorithm for large databases. In: Int. Conf. on Management of Data, pp. 73–84 (1998)

86. Guha, S., Rastogi, R., Shim, K.: ROCK: A robust clustering algorithm for categorical attributes. Information Systems 25(5), 345–366 (2000)

87. Gut, A.: Probability: A Graduate Course. Springer (2005)

88. Hall, E., Wise, G.: On optimal estimation with respect to a large family of cost functions. IEEE Trans. Information Theory 37(3), 691–693 (1991)

89. Hampshire II, J., Pearlmutter, B.: Equivalence proofs for multilayer perceptron classifiers and the Bayesian discriminant function. In: Touretzky, D., Elman, J., Sejnowsky, T., Hinton, G. (eds.) Proc. 1990 Connectionist Models Summer School. Morgan Kaufmann, San Mateo (1990)

90. Hampshire II, J., Waibel, A.: A novel objective function for improved phoneme recognition using time-delay neural networks. IEEE Trans. on Neural Networks 1(2), 216–228 (1990)

91. Harrison, M.: Introduction to Formal Language Theory. Addison-Wesley Series in Computer Science (1978)

92. Hartuv, E., Schmitt, A., Lange, J., Meier-Ewert, S., Lehrachs, H., Shamir, R.: An algorithm for clustering cDNAs for gene expression analysis. In: Annual Conf. on Research in Computational Molecular Biology, pp. 188–197 (1999)

93. Hartuv, E., Shamir, R.: A clustering algorithm based on graph connectivity. Information Processing Letters 76(4-6), 175–181 (2000)

94. Hastie, T., Tibshirani, R., Friedman, J.: The Elements of Statistical Learning. Springer (2001)

95. Haykin, S.: Neural Networks: A Comprehensive Foundation, 2nd edn. Prentice-Hall, New Jersey (1999)

96. Haykin, S.: Neural Networks and Learning Machines, 3rd edn. Prentice-Hall (2009)

97. He, Z., Xu, X., Deng, S.: K-ANMI: A mutual information based clustering algorithm for categorical data. Computing Research Repository abs/cs/051 (2005)

98. Hero, A., Ma, B., Michel, O., Gorman, J.: Applications of entropic spanning graphs. IEEE Signal Processing Magazine 19(5), 85–95 (2002)

99. Hild II, K., Erdogmus, D., Principe, J.: Blind source separation using Rényi's mutual information. IEEE Signal Processing Letters 8, 174–176 (2001)

100. Hirose, A.: Complex-Valued Neural Networks. Springer (2006)

101. Hirose, A.: Complex-valued neural networks: The merits and their origins. In: Int. Joint Conf. on Neural Networks, pp. 1209–1216 (2009)

102. Hochberg, Y., Tamhane, A.: Multiple Comparison Procedures. John Wiley & Sons, Inc. (1987)

103. Hochreiter, S., Schmidhuber, J.: Long short-term memory. Neural Computation 9(8), 1735–1780 (1997)

104. Hsin, H., Li, C., Sun, M., Sclabassi, R.: An adaptive training algorithm for back-propagation neural networks. IEEE Trans. on Systems, Man, and Cybernetics 25, 512–514 (1995)

105. Huang, G.B., Babri, H.: Upper bounds on the number of hidden neurons in feedforward networks with arbitrary bounded nonlinear activation functions. IEEE Trans. on Neural Networks 9(1), 224–229 (1998)

106. Huber, P.: Robust estimation of a location parameter. The Annals of Mathematical Statistics 35(1), 73–101 (1964)

107. Hubert, L., Arabie, P.: Comparing partitions. Journal of Classification 2(1), 193–218 (1985)

108. Hyafil, L., Rivest, R.L.: Constructing Optimal Binary Decision Trees is NP-complete. Information Processing Letters 5(1), 15–17 (1976)

109. Jacobs, R.: Increased rates of convergence through learning rate adaptation. Neural Networks 1, 295–307 (1988)

110. Jacobs, R., Jordan, M., Nowlan, S., Hinton, G.: Adaptive mixtures of local experts. Neural Computation (3), 79–87 (1991)

111. Jacobs, R., Peng, F., Tanner, M.: A bayesian approach to model selection in hierarchical mixtures-of-experts architectures. Neural Networks 10(2), 231–241 (1997)

112. Jain, A., Dubes, R.: Algorithms for Clustering Data. Prentice Hall Int. (1988)

113. Jain, A., Murty, M., Flynn, P.: Data clustering: a review. ACM Computing Surveys 31(3), 264–323 (1999)

114. Jain, A., Topchy, A., Law, M., Buhmann, J.: Landscape of clustering algorithms. In: 17th Int. Conf. on Pattern Recognition, vol. 1, pp. 260–263 (2004)

115. Janzura, M., Koski, T., Otáhal, A.: Minimum entropy of error principle in estimation. Information Sciences (79), 123–144 (1994)

116. Jensen, D., Oates, T., Cohen, P.: Building simple models: A case study with decision trees. In: Liu, X., Cohen, P., Berthold, M. (eds.) IDA 1997. LNCS, vol. 1280, pp. 211–222. Springer, Heidelberg (1997)

117. Jenssen, R., Hild II, K., Erdogmus, D., Principe, J., Eltoft, T.: Clustering using Rényi's entropy. In: Int. Joint Conf. on Neural Networks, pp. 523–528 (2003)

118. Jeong, K.H., Liu, W., Han, S., Hasanbelliu, E., Principe, J.: The correntropy MACE filter. Pattern Recognition 42(5), 871–885 (2009)

119. Jirina Jr., M., Jirina, M.: Neural network classifier based on growing hyperspheres. Neural Network World 10(3), 417–428 (2000)

120. Johnson, E., Mehrotra, A., Nemhauser, G.: Min-cut clustering. Mathematical Programming 62, 133–151 (1993)

121. Johnson, N.L., Kotz, S., Blakrishnan, N.: Continuous Univariate Distributions, vol. 1. John Wiley & Sons (1994)

122. Johnson, N.L., Kotz, S., Blakrishnan, N.: Continuous Univariate Distributions, vol. 2. John Wiley & Sons (1995)

123. Jordan, M., Jacobs, R.: Hierarchical mixture of experts and the EM algorithm. Neural Computation 6, 181–214 (1994)

124. Kannan, R., Vempala, S., Vetta, A.: On clusterings: Good, bad, and spectral. In: Annual Symposium on the Foundation of Computer Science, pp. 367–380 (2000)

125. Kapur, J.: Maximum-Entropy Models in Science and Engineering. John Wiley & Sons, New York (1993)

126. Karypis, G.: Cluto: Software package for clustering high-dimensional datasets, Version 2.1.1 (2003)

127. Karypis, G., Han, E.H., Kumar, V.: Chameleon: Hierarchical clustering using dynamic modeling. IEEE Computer 32(8), 68–75 (1999)

128. Kaufman, L., Rousseeuw, P.: Finding Groups in Data: An Introduction to Cluster Analysis. John Wiley & Sons, New York (1990)

129. Kittler, J., Hatef, M., Duin, R., Matas, J.: On combining classifiers. IEEE Trans. Pattern Analysis and Machine Intelligence 20(3), 226–239 (1998)

130. Kulkarni, A.: On the mean accuracy of hierarchical classifiers. IEEE Trans. on Computers C-27(8), 771–776 (1978)
131. Kullback, S.: Information Theory and Statistics. Dover Pub. (1968)
132. Lagarias, J., Reeds, J., Wright, M., Wright, P.: Convergence properties of the Nelder-Mead simplex method in low dimensions. SIAM Journal of Optimization 9, 112–147 (1998)
133. Lazo, A.V., Rathie, P.N.: On the Entropy of Continuous Probability Distributions. IEEE Trans. on Information Theory 24(1), 120–122 (1978)
134. Lee, Y., Choi, S.: Minimum entropy, k-means, spectral clustering. In: IEEE Int. Joint Conf. on Neural Networks, vol. 1, pp. 117–122 (2004)
135. Lee, Y., Choi, S.: Maximum within-cluster association. Pattern Recognition Letters 26(10), 1412–1422 (2005)
136. Lewis, F.: Optimal Estimation. With an Introduction to Stochastic Control Theory. John Wiley & Sons (1986)
137. Li, H., Zhang, K., Jiang, T.: Minimum entropy clustering and applications to gene expression analysis. In: IEEE Computational Systems Bioinformatics Conf., pp. 142–151 (2004)
138. Li, J., Liu, H.: Ensembles of cascading trees. In: Int. Conf. Data Mining, pp. 585–588 (2003)
139. Li, R.H., Belford, G.: Instability of decision tree classification algorithms. In: Proc. of the 8th ACM SIGKDD Int. Conf. on Knowledge Discovery and Data Mining, KDD 2002, p. 570 (2002)
140. Linsker, R.: Self-organization in a perceptual network. IEEE Computer 21, 105–117 (1988)
141. Liu, W., Pokharel, P., Principe, J.: Correntropy: Properties and applications in non-Gaussian signal processing. IEEE Trans. on Signal Processing 55(11), 5286–5298 (2007)
142. Looney, C.: Pattern recognition using neural networks: theory and algorithms for engineers and scientists. Oxford University Press, New York (1997)
143. Mackey, M., Glass, L.: Oscillation and chaos in physiological control systems. Science 197, 287–289 (1977)
144. Maghsoodloo, S., Huang, C.: Computing probability integrals of a bivariate normal distribution. Interstat. (334) (1995)
145. Magoulas, G., Vrahatis, M., Androulakis, G.: Effective back-propagation with variable stepsize. Neural Networks 10, 69–82 (1997)
146. Magoulas, G., Vrahatis, M., Androulakis, G.: Improving the convergence of the backpropagation algorithm using learning rate adaptation methods. Neural Computation 11(7), 1769–1796 (1999)
147. Maimon, O., Rokach, L.: Decomposition Methodology for Knowledge Discovery and Data Mining: Theory and Applications. World Scientific Pub. Co. (2005)
148. Mandic, D., Goh, V.: Complex valued nonlinear adaptive filters. John Wiley & Sons (2009)
149. Marques de Sá, J.: Applied Statistics using SPSS, STATISTICA and MATLAB. Springer (2003)
150. Marques de Sá, J.: MSE, CE and MEE. Tech. Rep. 5/2008, Instituto de Engenharia Biomédica (PSI/NNIG) (2008)
151. Marques de Sá, J., Gama, J., Sebastião, R., Alexandre, L.A.: Decision Trees Using the Minimum Entropy-of-Error Principle. In: Jiang, X., Petkov, N. (eds.) CAIP 2009. LNCS, vol. 5702, pp. 799–807. Springer, Heidelberg (2009)
152. Marques de Sá, J., Sebastião, R., Gama, J.: Tree classifiers based on minimum error entropy decisions. Canadian Journal on Artificial Intelligence, Machine Learning and Pattern Recognition 2(3), 41–55 (2011)

153. Marques de Sá, J., Sebastião, R., Gama, J., Fontes, T.: New results on minimum error entropy decision trees. In: San Martin, C., Kim, S.-W. (eds.) CIARP 2011. LNCS, vol. 7042, pp. 355–362. Springer, Heidelberg (2011)

154. Matula, D.: Cluster analysis via graph theoretic techniques. In: Mullin, R., Reid, K., Roselle, D. (eds.) Proc. Luisiana Conf. on Combinatorics, Graph Theory and Computing, pp. 199–212 (1970)

155. Matula, D.: k-components, clusters and slicings in graphs. SIAM J. Appl. Math. 22(3), 459–480 (1972)

156. McCulloch, W., Pitts, W.: A logical calculus of the ideas immanent in nervous activity. Bulletin of Mathematical Biophysics 5, 115–133 (1943)

157. Meila, M., Shi, J.: A random walks view of spectral segmentation. In: 8th Int. Workshop on Artificial Intelligence and Statistics, pp. 8–11 (2001)

158. Mergel, V.: Test of goodness of fit for the inverse-Gaussian distribution. Math. Commun. 4(2), 191–195 (1999)

159. Mokkadem, A.: Estimation of the entropy and information of absolutely continuous random variables. IEEE Trans. Information Theory 35(1), 193–196 (1989)

160. Möller, M.: Efficient training of feed-forward neural networks. Ph.D. thesis, Computer Science Department, Aarhus University (1993)

161. Murthy, S.: Automatic construction of decision trees from data: A multidisciplinary survey. Data Mining and Knowledge Discovery 2(4), 345–389 (1998)

162. Nadarajah, S., Zografos, K.: Formulas for Rényi information and related measures for univariate distributions. Inf. Sci. 155(1-2), 119–138 (2003)

163. NCI60: Stanford NCI60 cancer microarray project, http://genome-www.stanford.edu/nci60/ (2000)

164. Neelakanta, P., Abusalah, S., Groff, D., Sudhakar, R., Park, J.: Csiszar's generalized error measures for gradient-descent-based optimizations in neural networks using the backpropagation algorithm. Connection Science 8(1), 79–114 (1996)

165. Nelder, J., Mead, R.: A simplex method for function minimization. Computer Journal 7, 308–313 (1965)

166. Ng, A., Jordan, M., Weiss, Y.: On spectral clustering: Analysis and an algorithm. In: Advances in Neural Information Processing Systems, vol. 14 (2001)

167. Ng, G., Wahab, A., Shi, D.: Entropy learning and relevance criteria for neural network pruning. Int. Journal of Neural Systems 13(5), 291–305 (2003)

168. Papoulis, A.: Probability, Random Variables and Stochastic Processes. McGraw-Hill Co. Inc. (1991)

169. Parzen, E.: On the estimation of a probability density function and the mode. Annals of Mathematical Statistics 33, 1065–1076 (1962)

170. Pelagotti, A., Piuri, V.: Entropic analysis and incremental synthesis of multilayered feedforward neural networks. Int. Journal of Neural Systems 8(5-6), 647–659 (1997)

171. Pérez-Ortiz, J., Gers, F., Eck, D., Schmidhuber, J.: Kalman filters improve LSTM network performance in problems unsolvable by traditional recurrent nets. Neural Networks 16(2), 241–250 (2003)

172. Pipberger, H., Arms, R., Stallmann, F.: Automatic screening of normal and abnormal electrocardiograms by means of digital electronic computer. In: Proc. Soc. Exp. Biol. Med., pp. 106–130 (1961)

173. Porter, M.: An algorithm for suffix stripping. Program 14(3), 130–137 (1980)

174. Principe, J.: Information Theoretic Learning: Rényi's Entropy and Kernel Perspectives. Springer (2010)

175. Principe, J., Xu, D.: Learning from examples with information theoretic criteria. Journal of VLSI Signal Processing Systems 26(1-2), 61–77 (2000)

176. Prokhorov, A.V.: Pearson Curves. In: Hazewinkel, M. (ed.) Encyclopaedia of Mathematics. Kluwer Academic Pub. (2002)

177. Quinlan, R.: C4.5: Programs for Machine Learning. Morgan Kaufmann Publishers, Inc. (1993)

178. Raileanu, L., Stoffel, K.: Theoretical comparison between the Gini index and information gain criteria. Annals of Mathematics and Artificial Intelligence 41, 77–93 (2004)

179. Rama, B., Jayashree, P., Jiwani, S.: A survey on clustering: Current status and challenging issues. Int. Journal on Computer Science and Engineering 2(9), 2976–2980 (2010)

180. Rand, W.: Objective criteria for the evaluation of clustering methods. Journal of the American Statistical Association 66, 846–850 (1971)

181. Rao, B.L.S.P.: Nonparametric Functional Estimation. Academic Press, Inc. (1983)

182. Raudys, S., Pikelis, V.: On dimesionality, sample size, classification error, and complexity of classification algorithm in pattern recognition. IEEE Trans. Pattern Analysis and Machine Intelligence 2(3), 242–252 (1980)

183. Rényi, A.: Probability Theory. Elsevier Science Pub. Co. Inc. (1970)

184. Reza, F.: An Introduction to Information Theory. Dover Pub. (1994)

185. Richard, M., Lippmann, R.: Neural network classifiers estimate Bayesian a posteriori probabilities. Neural Computation 3, 461–483 (1991)

186. Riedmiller, M., Braun, H.: A direct adaptive method for faster backpropagation learning: The Rprop algorithm. In: IEEE Int. Conf. on Neural Networks, pp. 586–591 (1993)

187. Robinson, A., Fallside, F.: The utility driven dynamic error propagation network. Tech. Rep. CUED/F-INFENG/TR.1, Cambridge University, Engineering Department (1987)

188. Rokach, L., Maimon, O.: Decision trees. In: Maimon, O., Rokach, L. (eds.) Data Mining and Knowledge Discovery Handbook. Springer (2005)

189. Rosasco, L., De Vito, E., Caponnetto, A., Piana, M.: Are loss functions all the same? Neural Computation 16(5), 1063–1076 (2004)

190. Rosenblatt, F.: The perceptron: a probabilistic model for information storage and organization in the brain. Psychological Review 65, 368–408 (1958)

191. Rueda, L.: A one-dimensional analysis for the probability of error of linear classifiers for normally distributed classes. Pattern Recognition 38(8), 1197–1207 (2005)

192. Rumelhart, D., Hinton, G., Williams, R.: Learning representations by backpropagation errors. Nature 323, 533–536 (1986)

193. Saerens, M., Latinne, P., Decaestecker, C.: Any reasonable cost function can be used for a posteriori probability approximation. IEEE Trans. on Neural Networks 13(5), 1204–1210 (2002)

194. Safavian, S., Landgrebe, D.: A survey of decision tree classifier methodology. IEEE Trans. on Systems Man and Cybernetics 21(3), 660–674 (1991)

195. Salzberg, S.: On comparing classifiers: Pitfalls to avoid and a recommended approach. Data Mining and Knowledge Discovery 1, 317–327 (1997)

196. Santamaría, I., Pokharel, P., Príncipe, J.: Generalized correlation function: definition, properties, and application to blind equalization. IEEE Trans. on Signal Processing 54(6-1), 2187–2197 (2006)

197. Santos, J.: Repository of data sets used on technical report Human Clustering on Bi-dimensional Data: An Assessment (2005), http://www.dema.isep.ipp.pt/~jms/datasets

198. Santos, J.: Data classification with neural networks and entropic criteria. Ph.D. thesis, University of Porto (2007)

199. Santos, J., Alexandre, L., Marques de Sá, J.: The error entropy minimization algorithm for neural network classification. In: Lofti, A. (ed.) Int. Conf. on Recent Advances in Soft Computing, pp. 92–97 (2004)
200. Santos, J., Alexandre, L., Marques de Sá, J.: Modular neural network task decomposition via entropic clustering. In: 6th Int. Conf. on Intelligent Systems Design and Applications, pp. 62–67. IEEE Computer Society Press (2006)
201. Santos, J., Marques de Sá, J.: Human clustering on bi-dimensional data: An assessment. Technical Report 1/2005, INEB - Instituto de Engenharia Biomédica, Porto, Portugal (2005)
202. Santos, J., Marques de Sá, J., Alexandre, L.: Batch-Sequential Algorithm for Neural Networks Trained with Entropic Criteria. In: Duch, W., Kacprzyk, J., Oja, E., Zadrożny, S. (eds.) ICANN 2005. LNCS, vol. 3697, pp. 91–96. Springer, Heidelberg (2005)
203. Santos, J., Marques de Sá, J., Alexandre, L.: Neural networks trained with the EEM algorithm: Tuning the smoothing parameter. WSEAS Trans. on Systems 4(4), 295–300 (2005)
204. Santos, J., Marques de Sá, J., Alexandre, L.: LEGClust - a clustering algorithm based on layered entropic subgraphs. IEEE Trans. on Pattern Analysis and Machine Intelligence 30(1), 62–75 (2008)
205. Santos, J., Marques de Sá, J., Alexandre, L., Sereno, F.: Optimization of the error entropy minimization algorithm for neural network classification. In: Dagli, C.H., Buczak, A., Enke, D., Embrechts, M., Ersoy, O. (eds.) Intelligent Engineering Systems Through Artificial Neural Networks, vol. 14, pp. 81–86. ASME Press (2004)
206. Savitha, R., Suresh, S., Sundararajan, N., Saratchandran, P.: A new learning algorithm with logarithmic performance index for complex-valued neural networks. Neurocomputing 72, 3771–3781 (2009)
207. Shawe-Taylor, J., Cristianini, N.: Kernel methods for pattern analysis. Cambridge University Press (2004)
208. Sherman, S.: Non-mean-square error criteria. IRE Trans. Information Theory 4(3), 125–126 (1958)
209. Shi, J., Malik, J.: Normalized cuts and image segmentation. IEEE Trans. on Pattern Analysis and Machine Intelligence 22(8), 888–905 (2000)
210. Silva, F., Almeida, L.: Speeding up backpropagation. In: Eckmiller, R. (ed.) Advanced Neural Computers, pp. 151–158 (1990)
211. Silva, J.E., Marques de Sá, J., Jossinet, J.: Classification of breast tissue by electrical impedance spectroscopy. Medical & Biological Engineering & Computing 38(1), 26–30 (2000)
212. Silva, L.M.: Neural networks with error-density risk functionals for data classification. Ph.D. thesis, University of Porto, Portugal (2008)
213. Silva, L.M., Alexandre, L.A., Marques de Sá, J.: Neural Network Classification: Maximizing Zero-Error Density. In: Singh, S., Singh, M., Apte, C., Perner, P. (eds.) ICAPR 2005. LNCS, vol. 3686, pp. 127–135. Springer, Heidelberg (2005)
214. Silva, L.M., Alexandre, L., Marques de Sá, J.: New developments of the Z-EDM algorithm. In: 6th Int. Conf. on Intelligent Systems Design and Applications, vol. 1, pp. 1067–1072. IEEE Computer Society Press (2006)
215. Silva, L.M., Embrechts, M.J., Santos, J.M., Marques de Sá, J.: The Influence of the Risk Functional in Data Classification with MLPs. In: Kůrková, V., Neruda, R., Koutník, J. (eds.) ICANN 2008, Part I. LNCS, vol. 5163, pp. 185–194. Springer, Heidelberg (2008)
216. Silva, L.M., Felgueiras, C., Alexandre, L., Marques de Sá, J.: Error entropy in classification problems: a univariate data analysis. Neural Computation 18(9), 2036–2061 (2006)

217. Silva, L.M., Marques de Sá, J., Alexandre, L.: Neural network classification using Shannon's entropy. In: European Symposium on Artificial Neural Networks, pp. 217–222 (2005)
218. Silva, L.M., Marques de Sá, J., Alexandre, L.: Data classification with multilayer perceptrons using a generalized error function. Neural Networks 21(9), 1302–1310 (2008)
219. Silva, L.M., Marques de Sá, J., Alexandre, L.: The MEE principle in data classification: A perceptron-based analysis. Neural Computation 22(10), 2698–2728 (2010)
220. Silva, L.M., Santos, J., Marques de Sá, J.: Classification performance of multilayer perceptrons with different risk functionals. Technical Report 1/2011, INEB - Instituto de Engenharia Biomédica, Porto, Portugal (2011)
221. Silverman, B.W.: Density Estimation for Statistics and Data Analysis, vol. 26. Chapman & Hall (1986)
222. Stam, A.: Some inequalities satisfied by the quantities of information of Fisher and Shannon. Information and Control 2, 101–112 (1959)
223. Stoller, D.: Univariate two-population distribution free discrimination. Journal of the American Statistical Association 49, 770–777 (1954)
224. Tapia, R.A., Thompson, J.R.: Nonparametric Probability Density Estimation. The John Hopkins University Press (1978)
225. Theodoridis, S., Koutroumbas, K.: Pattern Recognition. Academic Press (2009)
226. Tsai, H., Lee, S.: Entropy-based generation of supervised neural networks for classification of structured patterns. IEEE Trans. on Neural Networks 15(2), 283–297 (2004)
227. Vapnik, V.: Statistical Learning Theory. John Wiley & Sons (1998)
228. Vapnik, V.: The Nature of Statistical Learning Theory, 2nd edn. Springer (2010)
229. Verma, D., Meila, M.: A comparison of spectral clustering algorithms. Tech. Rep. UW-CSE-03-05-01, Washington University (2003)
230. Vidyasagar, M.: A Theory of Learning and Generalization, with Applications to Neural Networks and Control Systems. Springer (1997)
231. Vilalta, R., Achari, M., Eick, C.: Class decomposition via clustering: a new framework for low-variance classifiers. In: IEEE Int. Conf. on Data Mining, pp. 673–676 (2003)
232. Vogl, T., Mangis, J., Rigler, J., Zink, W., Alkon, D.: Accelerating the convergence of the back-propagation method. Biological Cybernetics 59(4-5), 257–263 (1988)
233. von Luxburg, U., Schölkopf, B.: Statistical learning theory: Models, concepts, and results. In: Gabbay, S.H.D., Woods, J. (eds.) Handbook for the History of Logic. Inductive Logic, vol. 10, Elsevier (2011)
234. Vretos, N., Solachildis, V., Pitas, I.: A mutual information based face clustering algorithm for movies. In: Int. Conf. on Multimedia and Expo, pp. 1013–1016. IEEE (2006)
235. Wand, M., Jones, M.: Kernel Smoothing. Chapman and Hall (1995)
236. Wang, X.F., Du, J.X., Zhang, G.J.: Recognition of Leaf Images based on Shape Features using a Hypersphere Classifier. In: Huang, D.-S., Zhang, X.-P., Huang, G.-B. (eds.) ICIC 2005. LNCS, vol. 3644, pp. 87–96. Springer, Heidelberg (2005)
237. Watanabe, S.: Pattern recognition as a quest for minimum entropy. Pattern Recognition 13(5), 381–387 (1981)
238. Watanabe, S.: Pattern Recognition: Human and Mechanical. John Wiley & Sons, Inc. (1985)
239. Weisstein, E.W.: Pearson System (2010), http://mathworld.wolfram.com/PearsonSystem.html

240. Werbos, P.: Generalization of backpropagation with application to a recurrent gas market model. Neural Networks 1(4), 339–356 (1988)
241. Williams, R., Zipser, D.: A learning algorithm for continually running fully recurrent neural networks. Neural Computation 1, 270–280 (1989)
242. Williams, R., Zipser, D.: Gradient-based learning algorithms for recurrent networks and their computational complexity. In: Chauvin, Y., Rumelhart, D.E. (eds.) Back-propagation: Theory, Architectures and Applications. Erlbaum, Hillsdale (1995)
243. Woodward, W.A.: Approximation of Type IV Probability Integral. Master's thesis, Texas Tech University (1971)
244. Wu, Z., Leahy, R.: An optimal graph theoretic approach to data clustering: Theory and its application to image segmentation. IEEE Trans. on Pattern Analysis and Machine Learning 15(11), 1101–1113 (1993)
245. Xu, D.: Entropy and Information Potential for Neural Computation. Ph.D. thesis, University of Florida, USA (1999)
246. Xu, D., Principe, J.C.: Learning from Examples with Quadratic Mutual Information. In: IEEE Signal Processing Society Workshop Neural Networks for Signal Processing VIII, pp. 155–164 (1998)
247. Xu, R., Wunsch II, D.: Survey of clustering algorithms. IEEE Trans. on Neural Networks 16(3), 645–678 (2005)
248. Xu, X.: Dbscan, http://ifsc.ualr.edu/xwxu
249. Hochberg, Y., Tamhane, A.C.: Multiple Comparison Procedures. John Wiley & Sons (1997)
250. Yuan, H., Xiong, F., Huai, X.: A method for estimating the number of hidden neurons in feed-forward neural networks based on information entropy. Computers and Electronics in Agriculture 40(1-3), 57–64 (2003)
251. Yuan, L., Kesavan, H.: Minimum entropy and information measure. IEEE Trans. Systems, Man, and Cybernetics 28(3), 488–491 (1998)
252. Zhang, G.: Neural networks for classification: a survey. IEEE Trans. on Systems, Man and Cybernetics, Part C (Applications and Reviews) 30(4), 451–462 (2000)
253. Zhang, T., Ramakrishnan, R., Livny, M.: BIRCH: An efficient clustering method for very large databases. In: ACM SIGMOD Workshop on Research Issues on Data Mining and Knowledge Discovery, Montreal, Canada, pp. 103–114 (1996)
254. Zhang, T., Ramakrishnan, R., Livny, M.: BIRCH: A new data clustering algorithm and its applications. Data Mining and Knowledge Discovery 1(2), 141–182 (1997)

Index

Symbols

1-of-c coding, 3, 4, 27, 142

A

Activation function, 59
Adaptive learning rate, 148
Adjusted Rand Index (ARI), 195
Algorithm complexity, 126, 150

B

Back-propagation, 140
   through time, 156, 161
Bandwidth, 42, 235
   optimal, 238
Batch mode, 149
Batch-sequential algorithm, 150
Bayes
   classifier, 7
   consistency, 7
   error, 16, 35, 94, 109
   linear discriminant, 74
   optimal error, 52
Bias, 50, 61
Binary criteria, 111

C

Case, 2
Circular uniform distribution, 62
Class
   label, 3
   prior, 7
Class conditional distribution, 7
Classification

   rule, 5
   supervised, 3
   unsupervised, 2
Classifier
   problem, 9
   training, 9
Clustering, 176
   hard, 176
   hierarchical, 176
   spectral, 177
Codomain restriction, 22, 30
Compound classifier, 80
Confusion matrix, 196
Consistent learning, 66, 83
Continuous output, 93
Convolution, 48
Correntropy, 132
   induced metric, 133
Covariance matrix, 50, 71
Critical points, 98, 117, 119
Cross-entropy, 19, 128, 133
Cross-error, 91
Cross-validation, 56, 155

D

Data
   classification, 1
   clustering, 2
   object, 2
   splitter, 93, 94, 122
Decision
   border, 17, 52, 59, 61, 82, 86
   surface, 4, 17
Decision tree, 3, 93, 110, 204
   C4.5 algorithm, 208
Dirac-δ, 26, 27, 46

comb, 27, 47
Discrete
    Entropy, 93
    Error variable, 93
    Output, 93
Discrete-output perceptron, 116
Dissimilarity measure, 179, 183
Distribution
    overlap, 67
    tanh-neuron, 225

E

Eigenvalue, 76
    matrix, 74
Empirical
    error, 6, 94
    estimate, 13
    MEE, 42, 51, 108
    risk, 8, 20
    SEE splits, 104
Entropic
    dissimilarity matrix, 183
    dissimilarity measure, 186
    proximity matrix, 184
Entropy
    conditional, 219, 220
    estimation, 241
    power, 220
    property, 32
    Rényi's, 29, 220
    Rényi's quadratic, 29, 51, 58, 99, 220,
        222, 241
    relative, 216
    Shannon's, 29, 51, 93, 100, 111, 219,
        221, 241
Entropy property, 38
Entropy-of-error criterion, 114
Epoch, 52
Error
    PDF, 52, 77
    rate, 56, 63
    test set, 56
    training set, 56, 147
Estimate
    empirical, 216
    resubstitution, 216
Exponential PDF, 31

F

Fat estimation, 49, 51, 60, 68, 78, 81,
        106, 126
Features
    categorical, 206

numerical, 206
Fisher information number, 244
Function
    down-saturating, 223
    indicator, 5
    kernel, 235
    loss, 5, 8, 9
    sigmoidal, 59
    strictly concave, 223
    thresholding, 4, 17, 59
    up-saturating, 32, 223
Functional, 13

G

Gaussian kernel, 42, 44, 78, 235, 242
Generalization, 6, 56, 127
Gini index, 111, 208
Gradient
    ascent, 44, 46, 126
    descent, 8, 9, 41, 126, 135
    weight function, 128, 134
Grammar
    $A^n B^n$, 168
    embedded Reber, 167
    Reber, 166
Graph theory, 177
Greedy algorithm, 8

H

Heaviside function, 22
Hessian, 75
    matrix, 119
Hidden neuron, 16
Hold-out method, 56

I

Impurity criteria, 111
Information gain criterion, 111, 113,
        208
Information potential, 27, 49, 71, 90
Instance, 2
Integrated mean square error, 43
Interval-end criterion, 108
Iterative optimization, 9, 18, 41

J

Joint distribution, 19

K

KDE, 104, 106, 124

Kernel
    bandwidth, 78, 125
    density estimate, 42
    smoothing, 43
kernel bandwidth, 238, 243
Kullback-Leibler divergence, 19, 29, 216

L

Learning
    algorithm, 4
    curve, 57, 63, 83
    machine, 6
    rate, 44
LEGClust, 178
Likelihood, 215
    log, 215
    maximum, 19, 215
Linear discriminant, 17, 74
Local minima, 149
Logistic sigmoid, 59, 69
Long short-term memory, 161

M

M-neighborhood, 180
Møller's risk, 134
Mahalanobis distance, 52
MEE perceptron, 63
Min-cut, 177
Minimum
    error entropy, 9
    probability of error, 6, 35
    risk, 8
Minkowski distance, 18, 179
MLP training, 142
Monotonic risk, 135
MSE, 98, 128, 133
MSE consistency, 243, 244
MSE-consistent, 237
Multilayer perceptron, 9
Mutually symmetric distributions, 102,
    108, 122

N

Nearest neighbors, 190
Nelder-Mead algorithm, 68, 70, 71, 85
Neural network, 3, 139
    complex-valued, 170
    modular, 199
    recurrent, 156
Normal equations, 9
NP complete, 8

O

Observed error, 6
Optimal
    decision rule, 94
    split point, 87, 88
Outlier, 18
    sensitivity, 130, 133
Over-fitting, 6

P

Parameter
    estimation, 9
    vector, 44
Parzen window, 235, 242
    estimate, 42, 49, 70
Parzen window estimate, 235, 243
Pearson system, 225
Perceptron, 126, 136
    discrete output, 93
    synoptic comparison, 79
Probability
    mass function, 93, 111
    posterior, 16
Probability of error, 4, 8, 35, 94
    criterion, 113

R

Rényi's quadratic error entropy, 27
Real time recurrent learning, 157, 161
Regression-like classifier, 4, 8, 9, 13
Regularization, 6, 152
Resubstitution estimate, 107
Resubstitution estimate, 13
Ring dataset, 82
Risk, 8
    EXP, 133, 140, 142
    functional, 13, 14, 27, 28
    ZED, 121, 128, 132, 133, 140, 142,
        157, 158
Rosenblatt's perceptron, 59

S

Saddle point, 76
SEE, 96, 107, 112, 116
Sequential mode, 149
Shannon's error entropy, 26, 96, 107,
    116
Similarity criteria, 3
Simple hold-out, 56
Smoothing parameter, 133, 145, 191

Soft-monotonic, 135
Split point, 86
Square-error, 14, 18
Squashing function, 59
Statistical test
    Dunn-Sidak, 210
    Friedman, 210
Stoller split, 117
    empirical, 94
    theoretical, 94
Subgraph, 188
Support vector machine, 35, 155

T

Target value, 3
Task decomposition, 199
Taylor expansion, 119
Test set, 56
Test set error, 52
Theoretical MEE, 42
Training set, 56
Training set error, 52
Tree leaf, 9
True error, 6

Turn-about value, 102, 109, 120
Twoing criterion, 112, 113, 208

U

Under-fitting, 6
Univariate split, 204, 205
Unweighed subgraph, 185

V

Vapnik-Chervonenkis distance, 58

W

Weight, 61
    vector, 50
Whitening transformation, 74, 75
Widrow-Hoff algorithm, 19

Z

Zero-Error
    density maximization, 121
    probability maximization, 121